The Environment, the Establishment, and the Law

The Environment, the Establishment, and the Law

HARMON HENKIN, MARTIN MERTA,
AND JAMES STAPLES

HOUGHTON MIFFLIN COMPANY · BOSTON

NEW YORK · ATLANTA · GENEVA, ILL. · DALLAS · PALO ALTO

632.95042

HEN

Printed in the U.S.A.

Library of Congress Catalog Card Number: 78-135004

ISBN: 0-395-11070-X

INTRODUCTION

As citizens of a democracy, we are taught in school that there are forces in government looking after our interests. It is implicit that there may be disagreement as to what the citizens' interests actually are, but once clearly defined, some branch of the governmental structure is supposed to serve as our St. George, slaying whatever dragons beset us.

For those of us who cherish America the beautiful and the natural life that graces our lands and waters, there have come to light some revealing shortcomings in the safeguards on which we supposedly depend. The present volume presents in fearsome detail the inadequate way in which the federal pesticides regulations protect the natural landscape from the avalanche of chemicals released upon it by the agricultural and chemical industries.

Of all the chemicals in question, none has attracted more attention or engendered more controversy than DDT. We recognize it now as a classic Jekyll and Hyde compound, capable of enormous benefit to agriculture and other human interests, and at the same time responsible for shocking damage to natural ecosystems. DDT was first released for general insecticidal application in 1947, after its use for military purposes in World War II. It was quickly and widely acclaimed as a wonder drug efficient in controlling insect pests and disease carriers. How could man forsee its insidious, long-range effects on the environment, its lethal effects on non-target species? A decade of widespread and rather indiscriminate use produced a disturbing number of cases of wildlife loss. The alarm was finally and forcefully sounded in 1962 by Rachel Carson in her book *Silent Spring*. By that time the compound and its derivatives had permeated living organisms throughout the lands and waters of the world. Five more years were required for scientists to document the damage, which proved to be far more extensive than even Miss Carson had foreseen. In addition to killing such sensitive species as robins and trout, DDT in minute amounts was found to be inhibiting the reproduction of many other vertebrates, and some shell fish and other invertebrates as well. The peregrine falcons of Europe and America were virtually disappearing because of DDT induced weakness in egg shells. Some populations of terns, grebes, and pelicans were similarly on the wane. In laboratories scientists found that coho salmon, successfully introduced into Lake Michigan, were spawning normally but the fry were dying when they assimilated the DDT saturated fat in their egg sacs. Today, the evidence of insidious wildlife damage continues to mount.

But as the magnitude of the problem has become clearer, the governmental mechanisms for dealing with it have appeared more ephemeral. Complaints by ecologists and demands for remedial action by conservationists have been "referred" from branch to branch of government. The profit motivated chemical industry has continued to defend the manufacture and sale of DDT; legislators representing agricultural communities were reluctant to ban an insecticide so widely used on the farm. The general public has been interested, but not to the point of pressing for action.

A forum which could bring the whole grisly business into the open was finally found in Wisconsin, when the Department of Natural Resources was asked to conduct a hearing on whether DDT was a pollutant of the waters of the state. The Wisconsin hearing was recognized by all interested parties as a major test of pesticide control, and government, industry, and conservation interests mobilized for the showdown.

The Environment, the Establishment, and the Law tells the story of this Wisconsin hearing, mostly through excerpts of the actual testimony. The text reveals how seriously DDT has contaminated the ecosystems of America, and for that matter, the ecosystems of the oceans and other continents. It reveals with equal clarity how poorly the existing governmental structures have served to protect us and the environment from such chemical affrontery. The Pesticides Regulation Division of the U.S. Department of Agriculture is presented in an especially poor light. If this division is the St. George in Washington who is to protect us from pesticides, we had best bare our own swords and join the conservationists in the defense of environment. The lawyer Yannacone who conducted the Wisconsin case against DDT is far more capable of killing dragons than the responsible federal agents who are employed to do so.

A. Starker Leopold
Berkeley, California

EDITOR'S NOTE

On October 28, 1968, the Citizens Natural Resources Association, Inc. filed a petition and on November 1, 1968 the Izaak Walton League of America, Inc., Wisconsin Division similarly filed a petition with the Department of Natural Resources of the state of Wisconsin requesting a declaratory ruling on whether DDT was an environmental pollutant within the definitions of Sections 144.01 (11) and 144.30 (9) of the Wisconsin Statutes, by contaminating and rendering unclean and impure the air, land, and waters of the state and making the same injurious to public health and deleterious to fish, bird, and animal life.

Hearings began on December 2, 1968 before Maurice H. Van Susteren, Hearing Examiner of the Wisconsin Department of Natural Resources and concluded on May 21, 1969. Examiner Van Susteren is equivalent to a judge of a court of general jurisdiction. His decisions are confirmed by the seven-man Board of Commissioners of the Department of Natural Resources. If confirmed, the ruling assumes the binding effect of a court order. It can be appealed to the Supreme Court of the state of Wisconsin.

This book was written under the senior authorship of Harmon Henkin with background material and editorial advice furnished by Martin Merta and James Staples.

CONTENTS

MADISON

With a Cast of Thousands

Before the show at Madison, Wisconsin was over, 32 persons ranging in occupation from politician, lawyer, and arborist to bureaucrat, medical doctor, and businessman had appeared to testify about DDT. Their knowledge—or lack of it—makes up the hearing transcript, a document that records some 2,500 pages of direct and cross-examination with a few thousand more pages of scientific, unscientific, and pictorial exhibits thrown in for good measure. Culling through that mass of transcript for the following chapters necessitated drastic cuts and deletions and the total omission of the testimony of half of the witnesses. These deletions and omissions reflect no objective value judgments on the quality of a man's work but instead reflect the general theme of this book: that events such as occurred in Madison must be seen in the larger context of the social and scientific climate in which they took place.

For instance, Dr. Robert Rudd wrote what perhaps is the best book ever on the effects of agricultural chemicals, *Pesticides and the Living Landscape*. Although stylistically the book is not as good as Rachel Carson's best-selling *Silent Spring*, many scientists consider Rudd's book to be a better chronicle of the ecological damage caused by pesticides. But Rudd's testimony is ignored, while the testimony of younger scientists like Charles Wurster and Robert Risebrough is detailed. The reason for this is simple. The Wursters have built on the foundation set up by the Rudds. They have fashioned from earlier work, the current attack against DDT.

But, in the tradition of the drama, it will perhaps aid the reader if the chapters which are directly taken from the transcript of the Madison hearings are preceded by a list of all who testified and by a brief synopsis of the main points of their testimony.

Patrick Buckley, a Wisconsin commercial arborist, testified that the arborists need DDT for Dutch elm disease control. However, research should be sponsored to find a substitute for the pesticide. Buckley seemed to feel in advance that DDT would be banned for Dutch elm disease control (which it was during the hearing) and that the arborists were unjustly being made into scapegoats when they used only half of the state's DDT.

Donald Chant, a biological control man from the University of Toronto, talked about the economic threshold as a concept in pest control.

R. Keith Chapman, a University of Wisconsin entomologist, showed slides which purported to prove that DDT gave miraculous results in controlling insects and improving the quality of produce. He opposed a ban on DDT.

Frank Cherms, a professor of poultry science at the University of Wisconsin, tried to cast doubt on the petitioner's thesis that chlorinated hydrocarbons were causing birds to produce thin eggshells by claiming that diseases and even sonic booms could "scare the shell out of birds."

Francis B. Coon was Director of the Wisconsin Alumni Research Foundation (WARF) Laboratories Pesticide Analysis Section. Coon testified that PCBs, plasticizing compounds, and DDT were confused sometimes during WARF laboratory analyses. It was a major surprise to many that Coon would be called in to testify for the DDT industry.

Paul De Bach, a biological control expert from the University of California, showed that DDT could make agricultural pest problems worse.

Isadore A. Fine, a University of Wisconsin economist, spoke about the economic importance of the tourist trade to Wisconsin. He stressed that outdoor recreation was vital to the state, putting some $600 million a year into Wisconsin's coffers. He did not mention DDT.

E. H. Fisher, the epitome of the squirt-gun entomologist, was Coordinator of Pesticide Use Education at the University of Wisconsin. Known to be one of the University's biggest backers of DDT, he was a violent critic of the pesticide's critics.

Theodore Goodfriend, a professor of medicine and researcher at the University of Wisconsin College of Medicine, was the only medical doctor called to testify for the petitioners.

William Gusey, a wildlife specialist presently with the Shell Chemical Company, talked about how, when he was employed by the federal government, there was much monitoring of pesticides and an awareness of what the effects of DDT were.

Wayland J. Hayes, the DDT industry's resident toxicologist, testified, as usual, to the non-toxicity of DDT. While working for the Public Health Service, Hayes had conducted experiments with humans on the toxicity or lack of toxicity of DDT and had defended the pesticide for years.

Harry W. Hays, head of the Pesticide Registration Division of the Department of Agriculture, told the story of how pesticides are and are not registered.

Joseph Hickey, a University of Wisconsin wildlife ecologist, described his worldwide investigation of the thin eggshell phenomenon and the population crash of the peregrine falcon.

Hugh Iltis, a botanist at the University of Wisconsin, told about the various botanical ecosystems of Wisconsin and how they were all interrelated.

S. Goran Lofroth, the Swedish scientist, came to talk about the worldwide contamination of mothers' milk with DDT residues.

Orie Loucks, a professor of botany at the University of Wisconsin, held a peculiar position at the hearings. He appeared early in the proceedings to define ecology and talk about the water and weather patterns as influences on the Wisconsin ecosystem, then appeared once more on the last day of the hearings to present a grandiose systems analysis model explaining the total impact of DDT on the Wisconsin regional aquatic ecosystem. Loucks' systems testimony was the first use of systems analysis in the courtroom, and according to one person who attended the hearing, "I hope it is the last."

Kenneth J. Macek, a Department of the Interior fisheries biologist, had done research which connected the reproductive failure of trout with environmental levels of DDT.

Louis A. McLean, attorney for the DDT industry during the first half of the Madison hearings, was called to the stand by the petitioners' lawyer Yannacone to establish for the record his long-standing position as intellectual hatchet man for the pesticide industry and to enter into that record his article about the sexual hangups of pesticide critics.

Lewis Mitness, a Wisconsin rural State Assemblyman, was a firm critic of DDT. He countered Patrick Buckley's testimony by saying that DDT was ineffective in Dutch elm disease control, a contention backed with figures comparing Wisconsin cities which had stopped and which had continued using the chemical.

Senator Gaylord Nelson, a staunch environmentalist and Senator from Wisconsin, was an almost obligatory first witness because of his long-time, well-publicized opposition to DDT. At Madison, he gave an impassioned speech against that pesticide, including within it a synopsis of some of the current scientific data on the subject.

Bailey Pepper, an entomologist from Rutgers University, tried to cast doubts on the feasibility of alternatives to pesticides in the control of insect pests. He cited efforts now in progress to perfect nonchemical alternatives but said that they were not completely successful at the present time. He was against the proposed DDT ban.

Taft A. Pierce, an official of an exterminating company with some 40 offices and 500 employees throughout the Midwest, testified that his agency used DDT primarily to control mice and bats.

Paul E. Porter, a chemist from the Shell Development Company, one of the larger manufacturers of pesticides other than DDT, described how DDT breaks down in the environment.

Robert Risebrough, a Berkeley molecular biologist, ornithologist, and marine ecologist, furnished key evidence about the effects of DDT and its metabolites on birds, especially on their reproduction. A world expert in gas chromatography, he could analytically distinguish between PCBs and DDT.

Samuel Rotrosen, President of Montrose Chemical Company, a large manufacturer of the pesticide DDT, talked about the exports of DDT to underdeveloped countries and how the amounts of DDT being manufactured for use in this country had declined over the last ten years.

Robert L. Rudd, a University of California zoologist specializing in animal populations, testified to the ways DDT can limit the population of a species and how this limitation affects the ecosystems involved.

Alan Steinbach, a neurophysiologist at the Albert Einstein College of Medicine, gave the history of neurophysiology and, with that background established, showed how DDT affects the nervous system at the molecular, biochemical, and physiological levels.

Lucille F. Stickel, Director of the Pesticide Research Section at the Department of the Interior Fish and Wildlife Laboratories, Patuxent, Maryland, had definitely proven the theory that DDT and its metabolites can cause certain birds to lay thinner-shelled eggs.

Robert van den Bosch, the delightful biological control specialist from the University of California at Berkeley, chronicled the history of DDT use and abuse in commercial agriculture and told of effective alternative methods of pest control.

George Wallace, the Michigan State ornithologist and principal critic of DDT during the 1950's, was called as a footnote to history and furnished the field evidence of nervous system damage to birds from DDT.

Richard W. Welch, a pharmacologist from the Burroughs Welcome Laboratories, had been involved in significant research concerning DDT and enzyme induction in rats.

Charles F. Wurster, Jr., the self-styled molecular ecologist from the State University of New York at Stony Brook, in essence outlined the case presented by the petitioners.

So, that is the cast of characters; the following chapters, in relatively brief form, make up the plot.

1

Why Madison?

From an historical perspective why should Madison, Wisconsin in 1968–1969 have been the setting for the first major environmental legal confrontation? What scientific, social, political, and legal factors converged in that university city of 130,000 to produce a dramatic five-month struggle over the most widely used of all pesticides, DDT? Why did the petitioners, the chemical industry, and the government agencies choose Madison as their battleground?

The simplest explanation is the unique Wisconsin law which set up the ground rules for the battle. But that is like saying the existence of a ring and referee are the cause of a championship fight. A massive conflict over DDT had been building up for years. The administrative apparatus which pitted skilled lawyers against each other in a quasi-judicial setting was just one aspect of what happened in Madison. As significant as the legal maneuverings attempting to prove the guilt or innocence of the defendant, DDT, was the context of the battle.

In many ways the hearings mirrored in microcosm the intense social turmoil going on in this country today. At Madison the conflict between those longing to retain the status quo and those demanding change and responsiveness came bursting to the surface, revealing in the process the tactics of change as well as the need for it.

The law which overtly gave rise to the hearings and acted as an umbrella for the convoluted proceedings which followed, was deceptively simple. In Wisconsin, a state agency which enforces a law must hold public hearings if it is petitioned for a ruling as to the applicability of that law to a given existing situation. Under this statute, the Department of Natural Resources was petitioned by the Wisconsin Izaak Walton League and the Citizens Natural Resources Association of Wisconsin to judge whether DDT was a pollutant under Wisconsin Statute 144.01 which defines pollution to include "contaminating or rendering unclean or impure the waters of the state, or making the same injurious to public health, harmful for commercial or recreational use, or deleterious to fish, bird, animal or plant life."

But the hearings went far beyond the waters of Wisconsin and the Wisconsin regional ecosystem to critically and completely examine what DDT was doing to the earth's biosphere. Maurice Van Susteren, Senior Hearing Examiner for the Department of Natural Resources, realized the significance of the proceedings which he oversaw; he realized that they were not a run-of-the-mill examination of a simple problem and allowed them to become a forum to consider almost every aspect of the pesticide question in great detail. Thus, the enormous amount of information contained in the hearing record undoubtedly will be of great value to ecologists for years to come.

The anti-DDT coalition which formed to support the Citizens Natural Resources Association in its Madison efforts was a mixed bag, for the most part made up of people who had been fighting against the pesticide in their own ways for a number of years. They had followed excitedly legal efforts to get a complete hearing on DDT in Suffolk County, Long Island and in Michigan, efforts which, although legally inconclusive, resulted in curbing DDT use. So, when the coalition discovered the Wisconsin law, and felt they had a natural stage for the big battle, the two men essential to both the Long Island and the Michigan actions, Victor Yannacone, Jr., the Patchogue attorney who first legally challenged the use of DDT and Charles F. Wurster, an assistant professor of biological sciences at the State University of New York at Stony Brook, were invited in to guide their efforts. With the legal machinery already existing, the entire grim picture of the effects of DDT could be presented without worrying about the traditional bugaboos, jurisdiction and standing, that had restricted legal action in the past. A $10,000 fund-raising drive for the fight began around the state, the scientific data began to be collected and organized in mammoth quantities, and excitement began mounting among those who had been itching to get the DDT industry up against the wall.

Wisconsin, despite that notable exception of the fifties Senator Joe McCarthy, has long been one of the more progressive states in the Union, and leading the state intellectually has been Madison. Since the days of Fighting Bob La Follette's populist political ideas in the 1890's, Wisconsin has harbored many socially involved people including its two current senators, William Proxmire and Gaylord Nelson. As well, the University of Wisconsin is one of the best state universities in the country, and though an irritant to many conservative upstaters, has been a spotlight to many socially aware students, a group more and more prevalent these days. Both the progressive political and progressive educational establishments of Wisconsin came together to help the hearing: the University supplied witnesses, researchers, and helpers for the petitioners; and Senator Nelson supplied his national prestige to the proceedings as the leadoff witness.

Senator Nelson had been concerned about the state of the environment for a long time, even before ecology became a socially chic issue. He had tried vainly in three sessions of the Senate to pass a bill which would have banned the interstate sale of DDT and had long advocated a rational program for the general control of pesticides, but he hadn't gotten very far in his efforts, even in Wisconsin. Pesticides are intrinsically political in production and application. The Republican governor of the state, Warren Knowles, depended on agricultural support for political power. And whether the urban types who make up the general constituency of conservation organizations accept it or not, rural people do not like city folk butting in, telling them how to run their farming business. To many farmers, attempts to dictate what pesticides they may use on their crops are rank interference by dilettantes—and this is a factor not generally taken into consideration by the ecological establishment when it discusses agriculturally caused pollution. This issue of the needs and wishes of rural vs. urban dwellers was a factor within the hearings. The practical business of controlling agricultural pests was pitted many times against the more abstract issue of bird reproduction and fish hatching success. With this factor considered, it is no wonder that Governor Knowles did not back attempts to ban DDT.

Pesticides have been a major source of controversy since the publication of Rachel Carson's *Silent Spring* in 1962. Though her best-selling book chronicling the damages that pesticides were inflicting on the

The specific question before us is whether the overall benefits of DDT are offset by the damage it does. . . . This is a matter that must be measured in the long range, not the short. I think the evidence is clear that the damage is far bigger than the benefits.
Gaylord Nelson

environment spawned a national furor, congressional investigations, and a multitude of committees, the net result of the outcry was nothing as far as the tightening of pesticide control went.

There were many reasons why the pesticide problem did not begin to get solved at that time. One was the basic reluctance of scientists in the early 1960's to make statements in public for which they could be held professionally accountable. Scientists with anything resembling a social or political conscience were a rarity when *Silent Spring* appeared, and those who did speak out were considered to be some sort of freaks.

One rare outspoken scientist, George Wallace of Michigan State University, almost lost his job when he stated in front of a congressional investigating committee in the late 1950's that DDT was responsible for the widespread robin killoffs on the East Lansing campus. The DDT industry was too powerful a political force to tolerate much criticism from state employees and was capable of inflicting economic damage to its critics. But Wallace, who reappeared at Madison, was simply ahead of his time, and was fully vindicated, like Rachel Carson herself, when the scientific herd caught up with him.

Many scientists who loudly applauded the efforts of the petitioners in Madison had, five years before, been extremely leery of Miss Carson's work. In fact, some establishment-type specialists who became "radical" enough to attack DDT in 1969, had been among the more vociferous critics of *Silent Spring*, claiming that Miss Carson's data was "insufficient." Granted, much of *Silent Spring*, though amazingly prophetic, was not filled with enough "hard facts" and statistics to satisfy scientists who would not testify to their name in court without birth certificate in hand. But data firmly establishing the damage from DDT was amassed during the six years between *Silent Spring* and the Madison hearings.

The evidence which was accumulated by scientists around the world covered every property of DDT from its modes of travel in the biosphere; the subtle ways it affected the eggs of birds; the damage it could cause to fish reproduction; and the way the persistent pesticide affected nerve cells; to its potential for human damage. So much scientific information was assembled that all but the most intellectually insulated squirt gun *entomologists** and chemical industry flack scientists were admitting that the pesticide was causing tremendous damage, if not actually threatening a worldwide catastrophe of a major proportion.

The times and social mood of the country had also changed a great deal during the six-year period between *Silent Spring* and the Madison hearings. Movements for social change, such as civil rights and the

---

*Definitions of all subsequent italicized terms will appear in the glossary.

vicious internal debate raging over Vietnam, were producing a generation of activist types not satisfied with committees and hearings
ending with mere dead-ended reports. They wanted action, and to
some the legal approach finally became appealing.

Many academic types were being forced to admit that there was
a connection between what went on in the universities and laboratories and what was going on in the world. Hordes of stirred-up
students picketing, striking, and taking over buildings had caused the
specialists to examine their own positions in the world and had forced
them to try to ascertain whether they were part of the problems of
the country, at least in their own academic areas, or part of their
solutions. This was met by some with great personal regret. Nonetheless, by 1968, it was no longer a blatant sin for academia to become
involved with the social issues of the day and, in fact, in many scientific circles the opposite was true. Professors, especially younger ones
at some schools, felt obligated to get involved, at least in a cursory
way, with what was happening to and in society.

A goodly number of scientific types had been traditionally involved
with conservation causes, but the Madison hearings were far more
than traditional conservation activities. The difference between old-
line conservationists and the coalition which formed against DDT in
Wisconsin was extremely significant; the coalition aligned groups
which, in the past, had stood miles apart socially and politically.

For the most part conservation as a social force, since its inception
in the latter part of the nineteenth century under such American
aristrocrats as Gifford Pinchot and Theodore Roosevelt, has been an
upper- and upper-middle-class phenomenon. These conservationists
were interested in saving selected species of animals and selected
geographic locales for their own pleasures. They believed they could
save choice sections of the biological world solely by their socio-
economic positions. The word lobbying was too dirty to be used.

But history proved them wrong. The gentlemanly tradition of conservation bombed out in the face of increasing pressure and increasing
pollution from an increasingly technological society. These gentlemen
had pursued—and to a large extent still pursue—behind-the-scenes
influence strong enough to mobilize sufficient public opinion to
accomplish their goals. Yet, since many of the more prominent conservation groups had directorates intimately tied to corporate sources
of economic power, such groups could never amount to much of a
dynamic force.

The government benignly and wisely granted conservation groups
tax-exempt status, but this was like a chastity belt designed to insure
political virginity. If the groups got too active in pursuing conservation
evildoers, their tax exemption could be lost, because the government
that giveth could taketh away.

The first major change, and one that was to affect indirectly conservation's involvement in Madison, occurred in 1966 when the super-

prestigious Sierra Club became embattled with the federal government over a proposed dam which would have flooded part of the Grand Canyon. This dam was backed by Arizona Congressman Morris Udall, whose brother was then, coincidentally, Secretary of the Interior Stewart Udall, a man who had a general and sometimes undeserved public reputation as a conservationist. This battle marked the embryonic beginnings of a change for conservation groups from elitist organizations to something resembling progressive social groups.

The Sierra Club, with a base in San Francisco and a membership of 35,000, began taking out full-page ads in major newspapers to oppose the dam. The Internal Revenue Service, undoubtedly under the prodding of congressmen favoring the boondoggle, threatened to remove the Sierra Club's tax-exempt status on the grounds that the organization was lobbying—a forbidden political activity. Although this threat was serious to a group depending on tax-exempt contributions for its existence, the club took out another full-page ad in the *New York Times*. In essence, the ad said that, even though the Sierra Club would lose its tax-exempt status for continuing to fight the dam, if it refused to accept its moral responsibility it would have no reason to exist.

The club has now been fighting to get back its tax status for a number of years. However, the group's membership has risen to over 60,000 nationally and the organization has become a symbol across the country of aggressive and successful conservation political action. It is hard to argue with success, and the national press coverage the Sierra Club received during its battle with the government, paling that received by even the most sincere anti-litter campaigns, made other conservation groups rather green with envy.

It was easy to see that the times they were a-changing. Aggression was becoming the key word for groups and individuals seeking social change, and revolution was becoming a fashionable cocktail party word. By the late 1960's the whole machinery of government and industry had become fair game for the public; attacks were being mounted from both the political left and right, and grave doubts were being expressed in many quarters about the very functionability of the society. Military spending, the decay of the cities, widespread destruction of the environment, and other basic issues of national survival were making for a very restive public not afraid to question the assumptions of its national institutions.

Perhaps because of this climate, other environmental action groups were being quickly formed. Concurrent with the Sierra Club's growing militancy was the growing activity and prominence of the Long Island based Environmental Defense Fund. Here was a group which caught the ecologically involved public's eye. It didn't mince words. "Sue the bastards," became a flashy phrase to a great many people, and the group began riding a wave of popularity. Here was a group that, at least ostensibly, seemed to get things done in an exciting way.

*If Mankind should survive his own poisons—nuclear,
chemical, and social—it may be recorded that the light
generated in Madison in December of 1968 is
responsible.*
The Capital Times, *December 2, 1968*

However, there still was a basic schism between the Sierra Club
type groups and the Environmental Defense Fund which could be
characterized by one word, "people." The Sierra Club basically concerned itself with conserving and protecting nature for aesthetic
reasons, while the Defense Fund used an ecological approach to
emphasize the human need for a healthy environment. People were
relatively useless to the Sierra Club and other conservation groups
except when giving perspective to the height of a stand of redwood
trees. The title of one of the club's beautiful pictorial books might
be used to symbolize the conservationist approach to the environment, *In Wildness Is the Preservation of the World.* A comparison with
"Sue the bastards," and its economic and political implications, makes
the difference obvious. However, there were many individuals involved with EDF's law suits who were old guard conservation types,
a factor which was to be important in the eventual break-up of the
group.*

EDF's approach, under the tutelage of its lawyer Victor Yannacone,
was unique. The attorney, who had worked for the NAACP, labor

---

*See Appendix, pages 213–215.

unions, and others seeking legal redress and social justice on a broad scale, evolved the concept of a fundamental *constitutional* right to a healthy environment, a right rooted in the Ninth Amendment to the Constitution. People were the prime ingredient to Yannacone, not by any stretch of the imagination a conservationist, and the damage being done by the polluters was basically important to him in relationship to the human community. Before Yannacone entered the legal environmental scene, law suits brought by conservation groups sought to recover either personal damages or else sought the relief of real or threatened environmental damage affecting an individual or small group of individuals.

But Yannacone brought the class action to environmental litigation and sued in behalf of all the people being damaged or threatened by environmental despoliation regardless of their financial interest. This idea of his was almost pure Jeffersonian democracy applied to legal environmental practice. People, regardless of their individual station in life, had certain inalienable rights—including the right to a decent environment. Yannacone felt that the Constitution offered ample protection to the people in this area, but only if its tenets could be brought from the realm of theory into the world of practical legal combat. This unique democratic approach to environmental law is Yannacone's contribution to ecology and to law.

Included in all of Yannacone's class action law suits is a phrase which reads (in the case of the Madison action), "This suit is brought individually and on behalf of all those entitled to full benefit, use, and enjoyment of the national natural resource treasure that is the Wisconsin Regional Ecosystem without degradation or diminution in value resulting from the use of the broad spectrum, persistent, chemical biocide DDT, and all those similarly situated, not only of this generation, but of those generations yet unborn."

There are many ways of saying, "All power to the people," and Yannacone developed a legal counterpart to the revolutionary cry to arms favored by the new left.

Under this battle cry, Yannacone brought together to aid in the Citizens Natural Resources Association's unholy crusade against DDT at Madison, if only tentatively and tenuously, sportsmen, bird watchers, and those concerned about the effect that a tainted environment has on the quality of human life. The petitioners in the Madison case made up an awkward—at times, unwieldy—coalition whose members ranged from the new left to the old right with every gradient between. Although ultra-affluent found common cause with poverty stricken students, the internecine political and social struggles which developed in the coalition during the hearings made it sometimes difficult to realize that the common foe was DDT. The amazing thing, according to some who watched the infighting between coalition members, was that the ad hoc group managed to stay together at all during the five-month fray. The bitter back-room maneuverings over tactics, finances, and demeanor which developed and deepened as the hearings became more involved threatened many times to end the proceedings. Witnesses, each a high-ranking specialist in his own field, often acted as superstars and demanded the kid glove treatment. The lofty attitude of such scientists as Charlie Wurster, Hugh Iltis, and others was such that clashes erupted at times over the most minor points. Then, too, the coalition hierarchy became more and more firmly defined and cliquish as the hearings dragged on and on. And Yannacone, despite efforts of some EDF trustees to curb him, remained indubitably the ringmaster, alternating between soothing and aggravating the frayed tempers and temperaments of the coalition as it dug in for the long bitter fight, which had its roots some 1,000 miles away in Suffolk County, Long Island, where Carol Yannacone, a group of scientists and their "country lawyer" had gathered two years before to stop the spraying of DDT for mosquito control.

2

The Scientists Go to Prep School

Where is the courtroom genius of Yannacone? The scientists who testified for the petitioners at Madison were extremely impressed by the ebullient lawyer's grasp of even the most complex scientific issues and were dazzled by Yannacone's devastating cross-examination of key industry witnesses. But perhaps the most singularly impressive aspect of Yannacone at Madison was seen not in the courtroom but outside it. There, each evening before a session of the hearing, he presided over the most excruciating prepping of witnesses imaginable; scientists who would be testifying in court the next day would be all but terrorized. In fact, Yannacone was far more brutal with his own witnesses during preparation than were the industry's attorneys in court.

As one scientist who squirmingly endured one of Yannacone's prep sessions said later, "After Vic was through with me that night I wasn't even sure about my own name, let alone the validity of my own research." But Yannacone's technique worked almost to perfection. None of the petitioner's 16 witnesses came out of cross-examination badly scarred, a claim that the DDT industry's witnesses frequently could not make.

Yannacone's prepping was reminiscent of the Marine Corps technique of smashing a recruit's ego and remolding it in a tougher image. Scientists came to Madison full of bravado and professional self-assurance. But to Yannacone these qualities did not automatically make them into valuable witnesses who could hold up under heavy cross-examination. For the most part, the witnesses had no previous baptism under fire. They had never stood up to anything more intellectually dangerous than a student's question or a challenging letter to the editor of a journal in which they had published. But the courtroom is a jungle compared to even today's classroom or laboratory, and a clever lawyer is not a knee-jerk respecter of the rank or prestige of an opponent's witness. The only way a trial lawyer professionally survives is by winning his case, and the only way he can win is by diminishing the effect of witnesses on the other side.

Yannacone fully understood this, and each night after a communal dinner with the petitioners' forces, he would put tomorrow's "star" on the hot seat. He would allow the star to wax eloquently about his specialty and about what he planned to say at the hearing the next day. Then he would tear him apart, ripping out speculations, moralizing, and conclusions that couldn't be supported by hard, first-hand data. Often the potential witness, petrified and unsure of himself for the first time since his undergraduate days, would passively take Yannacone's advice on what he should and could not honestly say under oath. It worked. None of Yannacone's witnesses made a serious error on the stand. They had been through Yannacone's prepping hell, and the purgatory of the hearing room was not really challenging in comparison.

But even though Yannacone completely dominated those long hours of prepping, things didn't always proceed sweetly during them. Tempers flared, voices shrieked, and more than one scientist threatened to pack up his papers and reputation and charge back to the sanctuary of his university or laboratory. But whenever things became too hot and heavy and resentment toward Yannacone rose to the danger level, other scientists would jump into the verbal fray to soothe ruffled feathers and remind the witness, once more, how necessary it was to present only first-hand concrete data.

This special feature of science in the courtroom was an important aspect of the case. Before the Madison hearing, scientific information had seldom been broadly utilized in a legal setting. For one thing, putting scientific answers and questions in legal terminology isn't easy. In many ways, it's a mixed metaphor.

Take the matter of proof. To a scientist in a laboratory, a problem has not been solved; an answer cannot be accurately given if an exception to it exists. A judge or jury looking at scientific evidence, however, would not be so strict and would only demand that a preponderance of the evidence support a particular hypothesis or point of view. Perhaps the best example of this difference between the standards of proof in "pure" science and "courtroom" science can be summed up by a New York court ruling* often cited by Yannacone.

> While scientific accuracy demands of the scientist or doctor proof of cause which approaches absolute certainty, the law requires only a reasonable certainty or probability shown by a preponderance of the evidence . . . Plaintiff's proof is not required to soar into the icy stratosphere of certainty. It is enough, earth bound and flat footed, if it merely tips the scale of more probable than not.

This paragraph points to a distinction that was to become increasingly important to Yannacone the lawyer and to the scientists involved

---

Zaepfel v. E. I. Du Pont de Nemours and Company, 284 App. Div. 693 (N.Y.).

in courtroom environmental battles. Was, in Yannacone's phrase, "a reasonable degree of scientific certainty" enough? Could a reputable scientist stake that precious reputation on anything short of what he thinks at that time to be absolute? (Though one may ask if a scientist ever *really* knows anything for sure.) And, if he were willing to stick his neck out, how far? How would he define "reasonable certainty?"

Another problem arose. The courtroom is not the laboratory, and the evidence produced by the environmental specialists became more than a simple recounting of the scientific data that ivory tower science is about. Intertwined with the numbers, statistics, and regression lines were personal beliefs and comparative values.

The definition under Wisconsin law of "pollution" itself contributed to the subjective air which entered the hearing and could not be divorced from the hard science. Under Wisconsin law a pollutant is a substance, released into the environment, which has deleterious results. On the surface, this definition could cover every substance released by man into his world, and a strict enforcement of this statute would have the de facto effect of stopping human life itself. Once accepting, then, that it is impossible to ban *all* pollutants, choosing to ban one particular one rather than another is a value judgment. A more practical definition of a pollutant as a substance whose deleterious aspects outweigh its meritorious aspects, further highlights the moral ambiguities of the term.

In reality, then, scientists testifying in a case like that in Madison in an attempt to convince a judge or examiner that a particular form of environmental degradation is bad, have themselves, after examining the scientific evidence, made a moral decision. Thus, the battle in Madison was a struggle of values first and foremost, with scientific data as ammunition and the hearing examiner, Maurice Van Susteren, as referee for the combatants.

The effect of this ethical courtroom dichotomy is illustrated by the fact that Yannacone and his cohorts had then never won a clear-cut courtroom decision, yet were instrumental in restricting the use of DDT in the United States. The evidence the group brought together in Madison provided an almost overwhelming case for banning the pesticide. However, federal action finally occurred not because a judge ruled or Congress legislated that DDT was detrimental to the environment, but because hearing publicity aroused the public to a degree that even *Silent Spring* had not been able to match. Perhaps this is the nub of the legal approach to environmental problems.

The theme of choice and value set so often during the hearings was evident in the testimony of the very first witness, Gaylord Nelson, who, like the petitioners, had made up his mind before the hearings began. "This hearing" he said, "affords an opportunity to take a significant step that may well have historic consequences. The specific question before us is whether the overall benefits of DDT are offset by the damage it does . . . . This is a matter that must be measured in the long range and not the short. I think the evidence is clear that the damage is far bigger than the benefits."

Having Senator Nelson as the first witness at the Madison hearings was in some ways like having the President of the United States throw out the first ball of the baseball season. It was an honorary task for the man who had long been the Senate's leading critic of the misuse of pesticides; it was almost a sacred obligation to the environmentalists, now that the biggest battle yet was about to begin on Nelson's own turf. But Senator Nelson's testimony was ritualistic in one sense. It was a litany he had given many times before, and Louis McLean, the Task Force's lawyer, would not dare to attack it any more than the player who caught the President's first pitch would make a crack about his throwing arm. It was a preview of coming attractions.

In the days that followed, it became apparent to observers, especially scientific ones, that the legal maneuverings that go on during

Willard S. Stafford

scientific testimony in a courtroom have nothing to do with science. Some of these maneuvers attempted to establish guilt by association, others strove to cloud the issues; both were tactics which appalled many scientists who appeared. But in Madison these techniques often backfired, basically because industry—at least initially—depended too heavily upon them.

The technique of personal attack was something Louis McLean, the DDT industry's lawyer, had become a master of in his years with the Velsicol Chemical Company. But, while he was concentrating on personalities, scientists around the country were amassing data. McLean proved too rigid to cope with this development, the concerted broad-spectrum attack of the environmentalists, and the change in public mood. This undoubtedly was the reason that he was replaced during the hearing by Willard Stafford, a lawyer more polished, delicate, and better able to cope with the coalition of petitioners. Yannacone admits that if Stafford, a top midwestern trial

> The anti-pesticide leader . . . can almost always be iden-
> tified by the numerous variant views he holds about
> regular foods, chlorination and fluoridation of water,
> vaccination, public health programs, animal experimen-
> tations, food additives, medicine, science, and the busi-
> ness community, or by his insistence that insecticides
> should be mistermed "biocides."
> Louis A. McLean

lawyer, had been on the scene from the beginning, and had handled the cross-examination of such witnesses as Charles Wurster (who spread himself thin in his lengthy testimony and cross-examination), the petitioner's case might have appeared shakier than it turned out to be.

McLean spent too much time with such witnesses as Wurster and Robert Risebrough in attempts to discredit them professionally; something which had worked on anti-DDT witnesses in the past. But public sentiment and scientific evidence made his attempts puerile. The more McLean examined, the more the scientists talked for the record, and the better the environmentalists' position became.

Yannacone too was not above the discrediting technique, which he used to better advantage on McLean himself. After Senator Nelson's speech, Yannacone called McLean as his initial target in an attempt to damage the DDT industry's position by discrediting its attorney, a man who had frequently and in an unwarranted manner stuck his nose into scientific matters.

In a 1967 issue of *BioScience*, a publication of the American Institute of Biological Sciences, McLean had written an article claiming some amazing things about the critics of pesticides.* To summarize, he had said that they were of the "compulsive" variety, concerned excessively with sexual potency, and were primarily composed of health nuts and/or food faddists. This article was read into the record along with the austere data of long-trained scientists, making it a well-varied 2,500 page document.† Goading McLean into reading into the record some of his ludicrous written statements had little to do with scientific evidence and DDT's harmful nature but it did make good newspaper copy and did make the industry look rather peculiar.

The issue of "my scientists being better than your scientists" also popped up in the final brief of the Industry Task Force for DDT for the National Agricultural Chemical Association. In concluding this document, the DDT Task Force said, "In evaluating the evidence given in this proceeding, a comparison must be made between opposing witnesses. Those testifying in opposition to the petition (Dr. Pepper is a good example) brought to this hearing a combination of scientific training, actual experience and integrity which must give their [industry's] testimony great weight. On the other side all too often the

---

*Louis A. McLean, "Pesticides and the environment," *BioScience* 17 (1967): 613–617.

†The magazine *BioScience* flared up again in the trial when Yannacone objected to McLean lumping it in the same category with *Science*, the weekly journal of the American Association for the Advancement of Science in which Wurster had published some of his papers. "You can't lump *BioScience*, which publishes McLean's drivel, with *Science* magazine," bellowed Yannacone during the hearing. Yet, when Wurster wrote up the Madison hearings for publication, his article was published in none other than *BioScience*, illustrating how strange the legal shennanigans which make up a court record are, if viewed in a broader context.

witnesses ranged far outside of their own areas of expertise to express opinions which they were totally unqualified to render."

Whose scientists were "better"? A very deep question. Seemingly, the public said the petitioner's experts were better; at any rate, the petitioners were able to garner a much more inclusive body of data than was industry. But the record speaks for itself, and in its direct and cross-examination the story is told better than in any summary. To understand the drama of Madison one must turn to the record and let it stand for itself as much as possible.

3

Young Man with a Gripe

Q: Dr. Wurster, will you please state your name and residence for the record?

A: Charles F. Wurster, Jr. I live on Crane Neck Road, Oldfield, New York.

Charles Wurster was the hearing's crucial witness, at the same time the strongest and weakest person to take the stand for the petitioners. Although he had appeared in court against DDT in cases on Long Island, in Michigan, and in Milwaukee, and had the reputation for being one of the nation's most knowledgeable experts on *chlorinated hydrocarbon pesticides,* most of his knowledge came from literature-searching. By allowing Wurster to cover the broad spectrum of topics relating to the effects of DDT, Yannacone got a chance to outline what would become his entire case. But, at the same time, Yannacone made his witness an inviting target for industry's attorney, Louis McLean, because Wurster testified, unlike most of the petitioner's other witnesses, outside the realm of his personal scientific experience.

Wurster's testimony did more than set Yannacone's scene. It was directly damaging to DDT's reputation, and not only that, it put down the framework for the interdisciplinary ecological approach to environmental problem solving alien to industry and government but fundamental to the environmentalists in Madison. Wurster attempted to show many, many times during his direct testimony and cross-examination that the old approach to pest control predicated on the insulated work of isolated specialists would not suffice in the pesticide crisis that many scientists envisioned. If Wurster's testimony held up, it would not only succeed in destroying DDT's reputation but would also wreak havoc on the DDT industry's scientific methodology.

Yannacone spent little time in establishing Wurster's credentials, other than getting into the record that he taught at the State University of New York at Stony Brook and that he had personally done research with DDT. McLean broke in to ask what Wurster's doctorate was in and was told, organic chemistry. Then Yannacone started the important questions.

Q: Dr. Wurster, would you summarize for the Hearing Examiner and the record the general properties of DDT, with particular reference to those properties which have given rise to the substance of this particular petition?

A: This is, in effect, the answer to the question, but I would like to start it by describing what I would like to call an ideal insecticide.

Ideally, an insecticide does essentially two things. First, its action is restricted to the target organism, the pest. It doesn't kill other insects. It doesn't damage other organisms that you are not interested in. It does not upset the whole ecological system but rather goes to the heart of the matter, namely, that particular pest, and it kills it or reduces its population to some substantially lower level; but it does not interfere with other systems. And secondly, its action [is] restricted to the place where you put it. It should not be such that it escapes from the site of application and exerts activity elsewhere.

Now this is an ideal insecticide. Unfortunately, we do not have such an insecticide; there is no such ideal insecticide.

There are, however, many insecticides. At this point we have hundreds of insecticides registered. Some of them are better than others. We do have some that have extremely low mammalian or vertebrate toxicity, for example. We have others that are very high. We have some with low stability, some with high stability.

We can, in some ways, rate these [insecticides by] whether they are good or bad . . . according to [the] definition of ideality that I have given.

Unfortunately, the chlorinated hydrocarbons fail completely or virtually completely on both counts [of the definition]. Their activity is not restricted to the pest species.

At this point McLean broke into the examination with the first of his many challenges to Wurster's expertise and credentials.

Q: Pardon me, Doctor. I'm going to object to this answer as being beyond the ken of the witness unless you first qualify him as having some experience in pesticides.

Yannacone then attempted to qualify his witness by a series of questions to Wurster about his experience in the chemical analysis and study of biological effects of DDT, but McLean bore in on Wurster before Yannacone got any further.

Q: Have you—I know of two articles that you have published, one in regard to a robin count and one in regard to DDT in the Bermuda petrel—but I ask if you have written (on the subject of pesticides) any other technical articles describing your studies with them or your evaluation of them.

Wurster answered, yes. Then Yannacone laid a firmer foundation with a question designed to show that Wurster really had a firm grasp of and familiarity with the biological effects of DDT.

Q: Now Doctor, so we get the record developed in order, would you please enumerate for us in your own words—do it any way you wish—what the physical and chemical properties of DDT are that contribute to its biological effect on non-target organisms.

A: In many ways DDT is quite unique. What I will say holds, in part, for a number of other chlorinated hydrocarbons, including *dieldrin, aldrin, endrin, heptachlor* and several others, but I will speak specifically about DDT.

The uniqueness of DDT is caused by the fact that it [shows] a combination of four major factors. If it didn't have each of those four, if it didn't have the combination of all four, then we would have a very different situation than we do. . . .

Q: Doctor, for the record, so that we avoid confusion and objections later on, would you just simply state what the four elements are before explaining them?

A: Number one: [DDT] has broad biological activity that is not restricted to the pest organism.

Number two: It has great chemical stability. It's persistent, in common usage.

Number three: It is surprisingly mobile, . . .

Number four: It has solubility characteristics such that it is relatively insoluble in water and soluble in lipid tissue. . . .

Q: Now, would you start with the first [factor] and explain them in detail?

A: Biological activity: I mentioned [in my definition of the] ideal insecticide . . . that we would like its action to be restricted to the target insect, whatever it happens to be. This, unfortunately with DDT, is not the case. Its activity is very broad. It will kill a variety of beneficial insects. It will kill bees; various predators that may, in fact, be preying on the pest itself. It will, in effect, upset the insect ecology any place that it's put.

Further, its activity is not restricted to insects at all, but includes the entire phylum Arthropoda, arthropods. In other words, it will damage, in one way or another, or kill crustaceans* of various kinds including shrimp, crabs, and lobsters, just to give a few examples.

It also has broad activity through a number of other animal phyla. This would include, for example, such things as annelid worms. It's toxic to annelid worms. Further. . . . it's toxic to fish, birds, mammals, amphibians, and reptiles.

---

*Crustaceans are a class within the phylum Arthropoda. (Eds.)

Now this biological activity takes many forms. We have spoken
so far about [DDT's] tendency to kill an organism, but this is a
great oversimplification. It could affect an organism adversely
without killing it. . . .

At this point the hearing examiner, Maurice Van Susteren, cut in
and tried to figure out exactly what Wurster's specialty was, and was
told that, although his degree was in organic chemistry, his work
covered a variety of disciplines. This seemed to satisfy Van Susteren,
but Louis McLean, the DDT industry's attorney, was not so willing
to take Wurster's word.

Mr. McLean: Disciplines, specialities, I don't care how you call it. I
have not heard anything yet to qualify him in all these various
things. I know very few people—I have never met any person that
has all of these specialities, myself.

Mr. Yannacone: You have met one now. You're going to meet a few
more during this hearing. . . .

Yannacone continued the examination.

Q: Would you now continue with the discussion of the biological
activity of DDT?

A: I had mentioned that we have recently discovered that DDT has
estrogenic activity, that is, that it does function as a hormone,
a female sex hormone. We also know that is does so at incredibly
low concentrations. I don't know whether it was mentioned this
morning, but hormones function in the parts per billion range and
probably lower than that. So it is rather irrelevant to talk about
how tiny a part per billion or per million is when these compounds
are active at that level. It doesn't matter whether it's a jigger or
a drop [of vermouth] in a thousand-tank carload of . . . gin. This
is irrelevant. Because we do have extremely sensitive biological
systems, we have great biological activity at very low levels.

All right. Now let's go on to the second one of those points:
chemical stability.

We all know—I think we know—that DDT is a very stable
compound, that it has a *half-life* in the environment circulating
in world systems of at least a decade. I think that it's probably
a good bit more than that, but unfortunately, we don't really have
the data. Nobody has the data.

[In this connection] I will mention a paper by Nash and Woolson which was published in *Science* about 18 months ago. . . .*
This was a paper published by some people from the United States

---

*R. G. Nash and E. A. Woolson, "Persistence of chlorinated hydrocarbon insecticides
in soils," *Science*, 157 (1967): 924–927.

Department of Agriculture. [It] shows that 17 years after the application of DDT to a field in Maryland, something like 39% of the DDT was still there. The question of where the rest of it went was not treated in detail. It was hypothesized that some of it probably decomposed. And I'm sure, in fact, some of it did decompose. But where did the rest of it go? We [can't] really know the answer . . . unless we look where we would expect to find it. And that's what I will be getting into.

I do want to point out that analysis of, say, a treated soil or orchard that shows the presence of a residue after a certain number of years does not establish the half-life [of that residue] in the environment. It only establishes the length of time that some of it stayed where you put it. It is not logical—it's completely incorrect—to assume that all of the rest of it broke down, that it went away somewhere, [that] we can forget it. We are now

> *If DDT were a molecule with a high water solubility, we could afford to have a lot more of it around. We could lose it in those big oceans out there, because it would stay in the water, it would not come out of the water and go into living organisms.*
> *Charles Wurster*

learning that we can't forget it; that it goes [elsewhere] and turns up in all kinds of strange places.

All right. How does it get there? Let's go to point three: mobility.

If you look in the chemical and engineering handbook on DDT, [and check] its physical and chemical properties, your initial reaction would be: This is nice stable stuff; if you put it here, it's going to stay . . . and it won't cause any trouble; it won't get spread all over the place. This, it turns out, is altogether incorrect. DDT is much more mobile than we would ever have predicted. Why is this the case? There are a lot of reasons [which] I would like to go through . . . one by one. . . .

Point number one, under this subheading of mobility: DDT does have a water solubility. It's extremely low; it's one of the lowest of any organic chemical known. . . . The best estimate of its solubility is 1.2 parts per billion. . . . But it is not insignificantly low, because there is a fantastic amount of water on this earth. And so, if you have enough leaching, you have enough rain, you have enough water circulating around, water can, in solution, carry levels of DDT that are not insignificant. . . . In other words, even though, at any one time, if you look at water, you find its [DDT content is] either below the limits of detectability or it's virtually absurdly low, . . . there's so much water circulating around that the fact that there is [any] water solubility is a significant factor.

Now as far as transport in water is concerned, you would theoretically, on [the basis of] what I have said, say 1.2 parts per billion is all that water can carry. But this is not correct. Water can carry a good deal more, for two . . . reasons: . . . DDT has a very strong tendency to form *suspensions.* So, rather than being in *solution,* DDT can circulate suspended in water in much larger amounts. . . .

DDT [also] has a tendency to *adsorb* to particulate matter. This means if [DDT] is placed in some soil and then the soil erodes and is carried into streams, rivers, and into lakes, [the water] carries the particulates and the DDT [along with them]. . . .

An example of how dramatic this effect can be was recently shown in California in a paper by Keith and Hunt. . . .* When they took water from a number of locations in California and filtered . . . out the particulate matter; [they found that] the particulates . . . had between 10,000 and 100,000 times the concentration of DDT that was found in the water itself.

So water . . . can carry quite a bit of DDT, most of it not in solution.

*J. O. Keith and E. G. Hunt, "Levels of insecticide residues in fish and wildlife in California," *Transcript of 31st North American Wild Natural Resources Conference* (March 1966): 150–177.

Now I would like to move on and talk about the air. How does [DDT] get into the air?

... DDT does have a finite vapor pressure, [although] it's very low. [DDT is] a relatively nonvolatile material, so you would not expect much loss by volatility. But you do get some loss, because there is some vapor pressure.

But ... DDT, like its tendency to suspend in the water, also suspends in the air. Most methods of application, at least many methods of application, do their utmost to get the finest particles [possible; in other words,] a spray. The instrument used to spray elm trees usually is a mist blower. This sends a column of mist and very, very tiny droplets up into the air. A good bit of that material will not return to the ground or land on the tree; but the vehicle, the solvent, will evaporate and [leave] tiny crystals. These crystals can be carried into the atmosphere very, very great distances. They can travel, in that sense, all over the world.

Further evidence of this was shown by Dr. George Woodwell a number of years ago in Maine, when he measured the amount of DDT that reached the ground following spraying by an airplane.* When he measured the amount on the ground, he found that it was only half as much, approximately, as was released by the airplane. The rest of it did not reach the ground, but went elsewhere.

Now ... DDT [as well as adsorbing] to particulate matter and [being] carried by eroded particles into water. . . . is [also] adsorbed to particulate matter that is picked up by the wind and blown as dust around the world. This has been shown in a number of cases. In one of the cases, Dr. Robert Risebrough—sitting over there—found that in association with the dust over the island of Barbados there were dust residues of DDT.†

[It] has been shown, actually, in a number of cases, [that] DDT does adsorb with dust. This means when the wind blows over a treated field, it will pick up some of the material on that field, pick up some of the dust, and the DDT . . . with it.

Now [another] mechanism for movement is one that we have appreciated only rather recently. [It is] the phenomenon of co-distillation. Co-distillation means that when one substance passes into the vapor state, it carries another one along with it. In this case, when water goes into the atmosphere, it carries DDT along. . . .

Actually, this can be dramatically illustrated. Place on the counter here a beaker or glass of water that has suspended in it

*G. M. Woodwell, "Persistence of DDT in a forest soil," *Forest Science* 7 (1961): 194–196.
†R. W. Risebrough, et al., "Pesticides: Trans-Atlantic movements in the northeast trades," *Science* 159 (1968): 1233–1236.

something in the neighborhood of 10 parts per billion DDT, and let it sit here at room temperature for 24 hours. By this time tomorrow the concentration of DDT in that beaker [will] be roughly half of what it is now; it [will] be down to about five parts per billion. This was shown by workers from the U.S. Department of Agriculture.

And now, finally, there [are] storage mechanisms for [moving DDT] about. It can be transported within the bodies of living organisms that are themselves mobile. This is probably not a major mechanism for transport. [However,] in some cases it's probably important, particularly, for example, in the case of marine birds.

Many of the circulation patterns in the world are east to west patterns, whereas the migration of birds is often north to south. So, . . . if we [consider] some of the very abundant ocean birds like the Wilson's petrel or the shearwater [which] have very long migration routes north to south, . . . they can carry a not insignificant amount of DDT from one pole to the other. Nevertheless, I think this movement within living organisms is a relatively minor point when weighed against the other [points].

So we have, then, eight mechanisms whereby DDT can be distributed about the earth. When put in any one place, it is not going to stay there, but it's going to go to another place. I think it's absolutely essential to realize this point, because it's clear that once [the DDT] molecule is outside, there is no possible way to control it. . . . This means that it doesn't matter who uses DDT, where they use it, how they use it, [or] for what reason. The only thing that matters is whether they use it at all. Once it's outside, all of these mechanisms go into operation, and so the DDT takes off.

My point here is to emphasize that there cannot be the controlled use of an uncontrollable compound. There is no possible way to do it. There is no use in talking about indiscriminate or discriminate use of DDT, or reading the label or not reading the label, or being an expert or not knowing what you are doing. The only thing that matters is whether you use it or not.

Now, at this point, theoretically viewed, DDT is about the earth. This is not just a theoretical idea. Virtually every place we look for DDT, we find it. And if we understand [the] mechanisms [of distribution], it becomes rather clear why we should find it in so many places where we would not originally have expected to find it.

Senator Nelson mentioned yesterday that it's in the penguins in the Antarctic. Well, here's a whole continent where it's never been used. Nobody has ever taken any DDT and lost it in the Antarctic, because the insect population isn't worth bothering about. And yet . . . the penguins, and the fish that the penguins feed on, and the food that the fish eat, . . . all show traces of DDT.

After a brief recess, Wurster continued.

Dr. Wurster: . . . Workers* . . . [have] examined rainwater in Britain, and they also examined the air. . . . In all cases—in virtually all cases—they found residues of DDT present. Further, they examined air over rainwater from a number of cities in Britain, one of them being in extreme northern Scotland. Here's a place [with no] DDT in the immediate vicinity. And yet there was just as much [DDT] there as [in locations directly] to the east . . . that had agricultural areas.

In other words, at this point DDT is part of the normal circulation patterns of the earth. It is not [only] where you put it, but is all over the place, apparently wherever you look for it. Several years ago, some workers in Pittsburgh examined the air over that city and [found DDT] there.† So we have pretty well established that fact that [DDT] is in the air, . . . all the time, wherever you look.

Another example of how this operates is shown in a paper by Cole and Frear and several other authors.‡ [These men analyzed] a forest in Pennsylvania that had never been treated with any

---

*K. R. Tarrant and J. O'G. Tatton, "Organochlorine pesticides in rainwater in the British Isles," *Nature* 219 (1968): 725–727.

†P. Antomaria, M. Corn, and L. DeMaio, "Airborne particulates in Pittsburgh: Association with p,p'—DDT," *Science* 150 (1965): 1476–1477.

‡H. Cole, et al., "DDT levels in fish, streams, stream sediments, and soil before and after DDT aerial spray applications for fall cankerworms in Northern Pennsylvania," *Bulletin of Environmental Contamination and Toxicology* 2 (1967): 127–146.

chlorinated hydrocarbon insecticide. This was essentially a wilderness area that had no farms, no towns, no paved roads, . . . only forest and mountain countryside, and this was quite a substantial area. This area was to be treated with DDT for some insect that had come to the vicinity, [but] before [it was] treated with a chlorinated hydrocarbon insecticide, they analyzed the soil. They found the soil contained five parts per billion dieldrin, four parts per billion *DDE*, and nine parts per billion DDT. So how did it get there? Well, it obviously came down in the rain or came down as fallout or particulate matter.

Furthermore, they analyzed the trout in the streams. Those trout, . . . living in a watershed no part of which had ever been treated, . . . still contained .42 parts per million DDE, .10 parts per million *DDD,* 54 parts per million DDT, and .11 parts per million dieldrin. So . . . four different chlorinated hydrocarbon insecticides [were] present in the trout in those streams in a watershed that had never been treated at all. . . .

Examiner Van Susteren: Just a moment. The watershed had not been treated at all? What do you mean?

Dr. Wurster: [It] had never been sprayed; [they had] never sprayed the veins or any other part of the watershed of that stream. It had never been treated, either [with] DDT or any other chlorinated hydrocarbon. In other words, what was in that watershed, in the soil or in the fish, came down out of the sky.

Examiner Van Susteren: You mean to tell me [that neither] the farmers [nor anyone] else used [DDT] in the entire watershed?

Dr. Wurster: That's right. This is far up in the watershed; this is in the mountains. This isn't the entire river watershed. . . . This was 104,000 acres, . . . which is a pretty large forest.

Further, the suckers in the stream had a good bit higher residues. They had 2.4 parts per million DDE, 3.7 parts per million DDT, and 1.8 parts per million of dieldrin.

Now the water, before they treated this watershed, showed no DDT or any other chlorinated hydrocarbon. Then they began to spray in the vicinity [but] in a different watershed. . . . In other words, for several days they were spraying in surrounding watersheds. And the watershed I'm talking about showed DDT in the water before they began to treat [it].

Have I made that clear?

All right.

Now let me go to item number four.* I have now completed the section on mobility.

---

*You will recall that Wurster initially stated that he would discuss four characteristics of DDT: 1) its broad biological activity, 2) its persistence, 3) its mobility, and 4) its solubility. (Eds.)

Let me talk about the significance of the solubility characteristics of DDT.

If DDT were a molecule with a high water solubility, we could
afford to have a lot more of it around. We could lose it in those
big oceans out there, because it would stay in the water; it would
not come out of the water and go into living organisms. But it
is not soluble in water, and so it has a tendency to be picked
up by lipid tissue, which, in this case, is [found] in [all] living
organisms. . . . from the plankton right on up through the various
fish, birds, and so forth. All of these organisms have tissues that
can dissolve more DDT into them than can be dissolved in water.

Examiner Van Susteren: All right. Now for the benefit of, first, the
person who will study this record, "lipid" means fatty tissue?

Dr. Wurster: Lipid means fatty or fat-like. Lipids are usually defined
by their solubility characteristics. So a lipid is something that tends
to be soluble in a *nonpolar solvent* like hexane, or acetone.

Examiner Van Susteren: And it's l-i-p---

Dr. Wurster: L-i-p-i-d.

All living organisms contain lipids. Therefore, DDT is more
soluble in living organisms than it is in water. This means, then,
that all of the organisms in a body of water, whether it be a stream
or river, a lake or an ocean, . . . are busy scrubbing the DDT out
of the water and collecting it themselves. . . . This, then, explains
why, when we look at the water, we don't find [DDT]. . . . But if
we look at the organisms living in the water, there it is, sure
enough.

So we have, then, a situation whereby this material spreads
through the water [and] is constantly being taken out of the water
and contaminating things that live. This also indicates that if you
want to examine water quality, you don't examine the water, you
examine the organisms that live there, because you are always
going to be somewhere near the limits of detectability—the limits
of your instrument—if you are looking at water. . . . So this is not
the way to find out whether the water is of high quality. You
should, instead, look at the organisms that live there. . . .

Now let me go through [my] four points and point out the
significance of what this means. If we didn't have any one of these
four key points, we wouldn't have the unique problem that we
do with the chlorinated hydrocarbons.

Number one, the broad biological activity: If the [DDT] activity
were restricted to the pest, then we wouldn't have to worry about
non-target organisms, because if [DDT] reached them, it wouldn't
hurt them. But it does have this broad biological activity; and so
it can damage non-target organisms.

Secondly, if [DDT] were not stable, it would decompose before
it ever reached [non-target organisms]. If we put [DDT] on a field
of wheat or corn, by the time it had departed from the field, it

would have broken down into something non-toxic; and so, there again, it wouldn't reach many non-target organisms. But [DDT] is stable, and so it lives long enough to go a great distance from where you put it [to] contaminate non-target organisms a great distance away.

Thirdly, if [DDT] were not mobile, however toxic or stable it might be, it would stay where you put it, and, again, it would not reach non-target organisms. The fact that it is mobile means it can go away from where you put it and reach something else a long way off.

And finally, if [DDT] were not soluble in lipid tissue—and nonsoluble in water—we could spread it through the soils and bodies of water, and we wouldn't have to worry as much about it contaminating organisms. . . .

This morning we were talking about the organophosphates. Let's take a really bad actor like parathion. Here's an incredibly toxic material. . . . But it's really unstable, and in a matter of days or weeks it decomposes to relatively innocuous materials. It does not present the world problem that DDT does simply because it does not have one of those four characteristics; it doesn't have the stability, even though it's far more toxic than DDT ever was.

All right. Now there's one other thing that follows from what I said about the solubility characteristics that's also important to appreciate. And that is the phenomenon of *trophic level* concentration or biological concentration. Let us picture a *food chain.* . . . We will have, flowing out of the inorganic environment into that food chain, a certain amount of DDT, and it will flow in at all levels [of the] chain.

Let's say there are two different species of fish, one feeding on the other. Both of them will become somewhat contaminated [by] the environment. But the big fish is busy eating the little fish. Or the robin is busy eating the earthworm. And the robin eats many

Introduced only a quarter-century ago and spectacularly
successful during World War II in controlling body lice
and therefore typhus, DDT quickly became a universal
weapon in agriculture and in public health campaigns
against disease-carriers. Not surprisingly, by this time
DDT has thoroughly permeated our environment. It is
found in the air of our cities, in wildlife all over North
America and in remote corners of the earth, even in
Adelie penguins and skua gulls (both carnivores) in the
Antarctic.
George Woodwell

earthworms. Or the big fish eats many little fish. The tissues of
the prey are excreted, but he doesn't excrete DDT; he keeps it,
because once again it's more soluble in him than in the material
that is excreted. And so the DDT [from] many of the little fish
becomes accumulated in that one big fish. . . . If you go out and
look . . . at the big fish and the little fish on which he feeds, you
will almost invariably—not invariably, but almost invariably—find
the big fish has a substantially higher concentration of DDT than
the little fish on which he feeds. And so, within a food chain,
[concentration] tends to build as you go up the food chain. At
the lower end of the food chain you often have low levels of DDT.
When you go to the next link, you may multiply [the concen-
tration] by two or three or ten or even a hundred. Now when
you go to the next link, you multiply it again. So we have got a
situation where food chains are having DDT fed into them at all
levels, but [all DDT is] being concentrated up the food chain at
the same time.

Let me give an example of how this works, a very simple example that occurred right here in Madison and [in] a number of other
cities in the country. Spray DDT on elm trees. Some of the DDT

lands on the soil beneath the elm trees, and the earthworms
[busily take] it out of the inorganic parts of the environment. So,
the earthworms become contaminated. The earthworms have a
fairly simple nervous system, and so the DDT may very well not
kill the earthworms. Along come ground-feeding birds like robins
and chipping sparrows and others. They feed on the soil organisms
including the earthworms, and they become contaminated. And
so we have a fairly simple food chain effect.

It's not always that simple: that's a very nice simple classic
example.

But let's talk about another one. This is some work that I did
myself on Long Island. . . .

Mr. Yannacone: Would you identify this paper for the record?

Dr. Wurster: It's a paper by George Woodwell, myself, and Peter
Isaacson in *Science*, Volume 156, page 821, 1967.

Examiner Van Susteren: Entitled?

A: Entitled "DDT residues in an East Coast estuary: A case of biological concentration of a persistent insecticide."

In analyzing for DDT residues we analyze[d] quite a number
of organisms, and we then arrange[d] in a table . . . the---

Examiner Van Susteren: That's on page?

A: Page 822.

---we arrange[d] in a table the organism analyses in the order
of increasing [DDT] concentration. And lo and behold, by doing
so, we, in effect, put the food chain in order. . . . I will read just
a few examples of how this works.

Examiner Van Susteren: Slowly.

A: Slowly.

The zooplankton, a small crustacean, has four hundredths, .04
parts per million of DDT. That's 40 parts per billion. The shrimp,
the larger crustacean, some of whom are presumably feeding on
those smaller ones, have four times as much, 0.16 parts per million
[DDT].

Let's go a bit further here to some small fish, the needlefish or
a pickerel. The pickerel had 1.33. The needlefish 2.07. You see,
we have moved the decimal point over three times already. We
have gone through three orders of magnitude.

Going further, we have the terns. The common tern is feeding
on small fish. The red-breasted merganser is a diving duck that
feeds on larger fish. The red-breasted merganser had 22.8 parts
per million of DDT residues. The double-crested cormorant had
26.4 parts per million.

So, in going from the water to the top of this food web, we
have a concentration effect of, approaching one million. So you
see how efficient this system of uptake is, in going from the
inorganic environment of exceedingly low concentration to important concentrations toward the top of the food chain.

Table 1. DDT residues (DDT + DDE + DDD) (*l*) in samples from Carmans River estuary and vicinity, Long Island, N.Y., in parts per million wet weight of the whole organism, with the proportions of DDT, DDE, and DDD expressed as a percentage of the total. Letters in parentheses designate replicate samples.

Sample	DDT residues (ppm)	Per cent of residue as		
		DDT	DDE	DDD
Water*	0.00005			
Plankton, mostly zooplankton	.040	25	75	Trace
Cladophora gracilis	.083	56	28	16
Shrimp*	.16	16	58	26
Opsanus tau, oyster toadfish (immature)*	.17	None	100	Trace
Menidia menidia, Atlantic silverside*	.23	17	48	35
Crickets*	.23	62	19	19
Nassarius obsoletus, mud snail*	.26	18	39	43
Gasterosteus aculeatus, threespine stickle-back*	.26	24	51	25
Anguilla rostrata, American eel (immature)*	.28	29	43	28
Flying insects, mostly Diptera*	.30	16	44	40
Spartina patens, shoots	.33	58	26	16
Mercenaria mercenaria, hard clam*	.42	71	17	12
Cyprinodon variegatus, sheepshead minnow*	.94	12	20	68
Anas rubripes, black duck	1.07	43	46	11
Fundulus heteroclitus, mummichog*	1.24	58	18	24
Paralichthys dentatus, summer flounder†	1.28	28	44	28
Esox niger, chain pickerel	1.33	34	26	40
Larus argentatus, herring gull, brain (d)	1.48	24	61	15
Strongylura marina, Atlantic needlefish	2.07	21	28	51
Spartina patens, roots	2.80	31	57	12
Sterna hirundo, common tern (a)	3.15	17	67	16
Sterna hirundo, common tern (b)	3.42	21	58	21
Butorides-virescens, green heron (a) (immature, found dead)	3.51	20	57	23
Larus argentatus, herring gull (immature) (a)	3.52	18	73	9
Butorides virescens, green heron (b)	3.57	8	70	22
Larus argentatus, herring gull, brain‡ (e)	4.56	22	67	11
Sterna albifrons, least tern (a)	4.75	14	71	15
Sterna hirundo, common tern (c)	5.17	17	55	28
Larus argentatus, herring gull (immature) (b)	5.43	18	71	11
Larus argentatus, herring gull (immature) (c)	5.53	25	62	13
Sterna albifrons, least tern (b)	6.40	17	68	15
Sterna hirundo, common tern (five abandoned eggs)	7.13	23	50	27
Larus argentatus, herring gull (d)	7.53	19	70	11
Larus argentatus, herring gull‡ (e)	9.60	22	71	7
Pandion haliaetus, osprey (one abandoned egg)§	13.8	15	64	21
Larus argentatus, herring gull (f)	18.5	30	56	14
Mergus serrator, red-breasted merganser (1964)†	22.8	28	65	7
Phaldcrocorax curitus, double-crested cormorant (immature)	26.4	12	75	13
Larus delawarensis, ring-billed gull (immature)	75.5	15	71	14

*Composite sample of more than one individual. †From Captree Island. 20 miles (32 km) WSW of study area. ‡Found moribund and emaciated, north shore of Long Island. §From Gardiners Island, Long Island.

George M. Woodwell, Charles F. Wurster, Jr. and Peter A. Isaacson, "DDT residues in an East Coast estuary: A case of biological concentration of a persistent insecticide," *Science* 156(1967): 822. By permission.

Examiner Van Susteren: Now wouldn't some of the lower forms have just a small amount of lipid tissue?

A: Per cell, no.

Examiner Van Susteren: You do this on a per cell basis?

A: There may, in some cases, be some more lipid toward the top, but I don't think you could make that as a general rule. The major factor here is not uptake from the environment so much as biological concentration as far as the carnivore at the top is concerned. This is a larger factor than the uptake from the water, because if you take a cell or take an organism like the small crustaceans, they are not concentrating [DDT] so much through the food chain, and their amount is very low as compared to the [organism] at the top which could possibly be taking that much out of the environment directly.

Examiner Van Susteren: But the lipid tissues would be less?

A: [In] the one at the bottom? Not necessarily. In some cases I suspect so; but not necessarily.

Examiner Van Susteren: But the study was done on a cell basis?

A: On a whole organism basis.

Examiner Van Susteren: On a whole organism---

A: Yes, but [the] parts per million [figure] is not absolute in the organism.

Examiner Van Susteren: [Yet the study] wasn't done on a cell by cell basis?

A: No. This is the more or less conventional way of analyzing for total contamination. [You take] the whole organism—take a fish—grind him up . . . and analyze the sample of that ground-up fish.

Examiner Van Susteren: So then what you are saying is it wouldn't make any difference if the whale had two feet of blubber or had one inch. You are saying the DDT is concentrated in the lipid tissue.

A: Yes. But blubber is not the only kind of lipid tissue.

Examiner Van Susteren: I realize that. But you said that DDT was concentrated in lipid tissue, and the amount of lipid tissue in the zooplankton . . . would certainly be less than in a duck.

A: Yes, but we are talking about parts per million. I'm talking about a gram of fish versus a gram of zooplankton. In other words, we are not talking about absolute size. Insofar as a whale is concerned, if he has enough blubber, he can store quite a bit of DDT on an absolute basis.

All right. Let me give one somewhat more esoteric situation to show the effect of not only distribution and mobility, but biological concentration.

In the middle of the Atlantic Ocean is a bird called a Bermuda petrel.* It's a very rare bird [with] only some hundred of them

---

*C. F. Wurster and D. B. Wingate, "DDT residues and declining reproduction in the Bermuda petrel," *Science* 159 (1968): 979–981.

in the world. [It] is a completely pelagic (oceanic) bird. [It] does not come to land except to breed; and when it does so, it breeds on some very small islands; they are hardly bigger than this room; they are essentially rocks. This is the only time the Bermuda petrel touches land. It does not come close to our coastline, and so it is in no way in contact with any agricultural area or any [DDT] treated area of any kind.

The only effective way this bird can accumulate important amounts of DDT is through its food chain. The bird feeds mainly on cephalopods (small squids), and probably feeds on some fish as well. But [the petrel] is probably the top link in perhaps a four-step food chain, but a wholly oceanic food chain, not a coastal one in any regard. . . .

This bird, by the analyses of six specimens—we would like to have more, but we can't very well take samples of a population that is bordering on extinction—averaged 6.4 parts per million [DDT] in its eggs and its dead chicks. These were not live specimens, they were dead ones. We did not want to take any live specimens.

The point of this is that it shows that DDT is an important contaminant or pollutant in the oceanic food chain at this point. There is clear evidence—not just [this], there is other evidence—that DDT is an important factor in the contamination of the oceans at this point.

That summarizes my prepared presentation.

Mr. Yannacone: All right, Doctor, is there anything else you might want to add about the physical and chemical properties of DDT as they relate to the impact of DDT on the environment?

A: I think that the DDT picture at this point has become extremely serious, much more than is generally realized. I think it has become a very substantial threat to a number of organisms, particularly among birds and fish. . . .

Mr. Yannacone: All right, Doctor, thank you.

Examiner Van Susteren: Mr. McLean?

Mr. McLean: Thank you, sir.

4

Wurster's Cross-examination

Louis McLean, billed in advance as the Task Force's hatchet man, seemed to epitomize to many people the DDT industry's waspish approach to any criticism of its products and techniques. And, true to that image, in cross-examining Wurster, McLean launched a stinging attack but, like a veteran club fighter just searching for an up-and-comer's weakspot, he first asked Wurster a few general questions.

Mr. McLean: Have you obtained some of your background information on the general subject of pesticides as a result of your association and work with the Environmental Defense Fund?
Dr. Wurster: Have I gotten the information from the Environmental———
Q: I say, have you gained this in connection with your association with the—part of your background?
A: I don't understand what you mean. You mean, has the Environmental Defense Fund been a part of my education?
Q: Have you gained this information in your activity for the Environmental Defense Fund?
A: Oh, yes, in contacting other scientists, definitely.

Then McLean threw his first kidney punch.

Q: Did you gain information on this subject in September of this year in Suffolk County, Long Island, when the Environmental Defense Fund filed an injunction suit in the courts there to enjoin the use of malathion for the control of mosquitoes———

What didn't make the hearing record in the uproar that ensued was the hooker at the end of McLean's question: he claimed that there was an encephalitis epidemic on Long Island at that time.

Mr. Yannacone: May I object at this time. I am counsel to the Environmental Defense Fund and as long as I have been with the Environmental Fund—since it was incorporated—the Environmental Defense Fund has never filed any suits on malathion. Now let's take that again, sir.

Mr. McLean: I will get into that in a minute.

Examiner Van Susteren: You want to complete your question and---

Mr. McLean: ---while an encephalitis---

Mr. Yannacone: I am going to object to that question---

Mr. McLean: May I complete my question?

Mr. Yannacone: ---unless Counsel can prove there was such a case brought by the Environmental Fund.

Examiner Van Susteren: Do you know?

Dr. Wurster: Yes, I can answer, no. . . . There was no such case brought by the Environmental Defense Fund. And I hope there isn't going to be and I do not get information from the---

In fact, as I recall, I was out of town. I was in Africa while that occurred; it came up while I was gone. I had nothing to do with it, nor did the Environmental Defense Fund.

Mr. McLean: Thank you. Do you know---

Mr. Yannacone: May I please, on the question, Mr. Examiner, on the question the witness was just asked---

Examiner Van Susteren: He just answered it.

Mr. Yannacone: All right. I'd be happy to submit a court record in that case.

Mr. McLean: The Environmental Defense Fund did not file this case either, but you are active as an organization for the petitioners, is that correct?

Dr. Wurster: Not entirely. That case was one where EDF---

Mr. Yannacone: I'm going to object. What are we talking about? The [case] here today---

Dr. Wurster: Both.

Mr. Yannacone:---or a case on Long Island? Or talking about a dieldrin case of last month?

Examiner Van Susteren: First of all, we should find out what connection, if any, Dr. Wurster has with EDF.

A: I am a member of the Board of Trustees of EDF and chairman of its scientist advisory committee.

Mr. McLean: And you were one of the original organizers of EDF, were you?

A: That's right.

Q: Do you know if EDF played any part in support of the lawsuit filed for an injunction against the use of malathion for mosquito control in Suffolk County, New York?

Mr. Yannacone: I'm going to object until he establishes there's been such a lawsuit.

Examiner Van Susteren: He's asking the witness if, in fact, a lawsuit has been started by the Fund. And certainly the witness would be qualified to know.

Dr. Wurster: EDF had no role whatsoever in the suit concerning malathion.

Its role here is, by formal resolution of the trustees, one of involvement. In other words, EDF is a part of this. In other words, we are not just sort of sneaking in the door and pretending not to be here. We are here.

Mr. McLean: All right.

Examiner Van Susteren: Well, the EDF is not one of the petitioners.

Dr. Wurster: That's right, we are here by invitation.

Mr. Yannacone: He filed an appearance.

Examiner Van Susteren: They filed an appearance as an intervenor.

Mr. McLean: Dr. Wurster, among your publications do you include a cartoon book, collaborating with several people including Dr. Woodwell, a cartoon book that was distributed gratuitously, but without invitation, at any of the science meetings in 1966? . . .

Now, Dr. Wurster, this cartoon book which is entitled *A-Bombs, Bugbombs, and Us* bears on the inside—I guess you'd call it actually the first page—G. M. Woodwell, W. M. Malcolm and R. H. Whittaker, Biology Department, Brookhaven National Laboratory, Upton, New York. You are not shown as a co-author at that point, are you?

A: Right.

Q: Are you shown as a co-author later in this book?

A: No.

And so it was to go through four days of cross-examination. McLean would alternate tacks: first the technical and scientific, then the political or social one. But this approach was to prove very dangerous for McLean. The harder he tried to nail Wurster, the more questions he asked. And the more questions he asked, the more information Wurster could pour into the record from his voluminous knowledge of the literature about chlorinated hydrocarbons.

But McLean persevered. In one segment of the cross-examination he tried to nail Wurster on the title he had given himself, "molecular ecologist."

Q: Let me ask you about that term. That's a new scientific term. I believe you said you had just conferred that degree on yourself. "Molecular," I believe, refers to something rather small. Isn't a molecule rather small?

A: Usually.

Q: And the ecosystem, which has the same [root] as ecologist, would be something rather big. This would conclude then, [if] my translation is correct, knowing a little about a great many things, kind of a jack-of-all-trades, is this it?

Examiner Van Susteren: Can you answer the question?

A: I don't know what to do with it.

I told somebody that I seem to be learning less and less about more and more and pretty soon I would know nothing about everything, but---

Cross-examination can be a very rough business. It gives prospective witnesses an incentive to be well prepared and confident of the validity of their testimony—or not testify at all.
Charles Wurster

Mr. Yannacone: Your Honor, I'm going to submit at this time that . . . we ought to give the witness or someone the opportunity to explain that there is in common usage today the phrase "molecular biologist," and it has a very definite meaning.

Examiner Van Susteren: He said "molecular ecologist."

Mr. Yannacone: And this happens to be the phrase "ecologist." This is the first time I have ever heard the combination of the two. I think we ought to give the witness the opportunity to fully explain them---

Dr. Wurster: Well all right---

Mr. Yannacone ---without the ridicule of Counsel.

Examiner Van Susteren: Counsel is not ridiculing. There are several people in the audience who are snickering, but they will be cautioned to keep their humor to themselves in this situation.

Could you explain the difference to us [between] molecular ecologist and molecular biologist?

A: Molecular biologist, in common usage, is often a molecular geneticist. My idea of a molecular ecologist is one familiar with the field of molecules or biochemistry with some capability in ecology; in other words, the ability to combine the two, or at least to try to do it. We have found, or we are finding, that biochemical systems have extremely important ecological consequences, so it's becoming vital to combine them, if not in one person, in many

people. But it's essential that there be interdisciplinary coopera-
tion and crossing of lines, because it's become extremely clear
that biochemistry plays a vital ecological environmental role.

Mr. McLean: I understand biochemistry—and I'm not trying to be cute
with you—but I would like to distinguish how you term yourself.
You do not, in calling yourself a molecular ecologist, consider
yourself as a medical toxicologist?

A: No.

Q: You have no medical training?

A: No.

Q: Nor do you consider yourself a plant pathologist?

A: No.

Q: I assume you took some courses in statistics, but you don't consider yourself a statistician?

A: I did not take any courses in statistics.

Q: You do not consider yourself an entomologist?

A: No.

Q: Nor an ornithologist?

A: I often consider myself as an ornithologist, yes.

Q: I assume you would consider yourself a biologist?

A: Yes, I am an assistant professor of biological sciences, so by
definition that makes me a biologist.

Q: Biology is a rather broad field that covers a lot of things and
touches very few in a great degree, just like ecology is . . . also a
very broad area, is that correct? And I believe you already said
you do not classify yourself as an expert in analytical chemistry?

Examiner Van Susteren: You said, yes?

A: Yes. But that doesn't mean I have never had anything to do with
analytical chemistry. It means I don't consider myself a specialist
in analytical chemistry.

Mr. McLean: Do you consider yourself as an expert even if not a
specialist?

A: Somebody else will have to judge whether I'm an expert in anything.

Pursuing his zig-zag course, McLean went after Wurster from a
different angle. Bringing up the subject that was, in many ways, the
key to the industry's defense of DDT, McLean suggested that polychlorinated biphenyls (industrial compounds, frequently found in the
environment) were interfering with analyses for DDT, thereby invali-
dating the research on which the anti-DDT forces based their case.
When this was sidestepped, McLean directed his examination to the
difficulties involved in—even the impossibility of—determining the
existence of DDT in an organism, when present in infinitesimal quantities.

Q: You referred to quite a bit of DDT residue data findings in your
direct testimony. What I wonder is, are the data giving the

amounts of DDT that you referred to irrevocably clear or did you have artifacts such as polychlorinated biphenyls (or PCBs).

A: I would like to refer questions on the polychlorinated biphenyls to Dr. Risebrough, who will be our next witness.

Q: Well isn't it certain that polychlorinated biphenyls are used as plasticizing agents?

A: That's correct.

Mr. Yannacone: I'm going to object. The witness has testified that he does not wish to talk about the polychlorinated biphenyls, and on direct examination he did not discuss the polychlorinated biphenyls. Unless Counsel is ready right now to show polychlorinated biphenyls have some relevancy, he can't ask the question.

Examiner Van Susteren: Let's give Mr. McLean a chance. Apparently there's some possibility of confusion in determination of levels as to whether they are DDT or DDD or polychlorinated---

Mr. McLean: In an examination for DDT, DDD, and DDE residues, may not polychlorinated biphenyls interfere as artifacts?

A: It depends which column you are using.* In some cases they may, in other cases they may not.

Q: And when you referred to your findings . . . and analyses, had you made a subtraction for the polychlorinated biphenyls? [Had you gone] through processes to eliminate them?

A: Yes.

Q: Would you describe how you eliminated them?

A: By confirming the identity by *thin-layer chromatography.*†

By this point, McLean was beginning to realize the opportunity that Wurster was getting to fill up the hearing record with evidence damaging to DDT. Yet, he continued. After considerable testimony on the accuracy of DDT detection equipment when used to measure varying samples, McLean asked:

Mr. McLean: What you call a peak "in the parts per billion range" [is a] pretty large peak?

A: When you are talking about a substance like ortho, para-DDT, which has been shown by the Department of Agriculture recently and by the Burroughs Wellcome Laboratory along the Hudson River in New York to have estrogenic activity and to have it at exceedingly low concentrations, then the parts per billion range is certainly one to give very serious thought about.

There was a recent estimate; in fact this was an estimate in the paper by Bitman and several other authors recently published in

---

*This statement refers to the processes of chromatography. For further explanation see the glossary, page 219. (Eds.)

†For a further discussion of this subject, see Dr. Risebrough's testimony, page 59 and the glossary.

*Science** (Bitman and these other authors are at the U.S.D.A. Laboratory in Beltsville, Maryland) . . . that in the world environment there was circulating something in the neighborhood of 200 million pounds of an estrogenic substance. And we have just discovered in the last year that DDT is an estrogenic substance.

Now this may to you seem to be insignificant. But to scientists this is not insignificant. This is a very alarming situation. . . .

It may be all right for you to say this is---

Mr. McLean: If the Court please---

Examiner Van Susteren: Now just a moment. We are getting into an argument.

Mr. McLean: Dr. Wurster, I don't want to get to the point of asking you just to give me a simple yes or no answer. But if you are going to give me extremely long and ranging answers to a question which I would hope to be answered rather simply, your answers are necessarily going to extend my cross-examination of you.

And I would appreciate it if you would---

A: When---

Mr. Yannacone: Your Honor, I respectfully request at this time that the Court instruct the witness that if the question cannot be answered yes or no, he need not answer yes or no.

Examiner Van Susteren: He may also ask for an opportunity to expand or amplify.

But as the questions continued to come from McLean and the answers voluminously from Wurster, the record grew fat with information about DDT and ecological problem solving.

Mr. McLean: Now Doctor, I believe when we adjourned yesterday, I was inquiring as to your familiarity with other factors that affect wild populations. Had we gotten to the botfly? Or are you familiar with any damage that the botfly does to wild populations?

A: Sir, I think I need to answer that at some length.

It seems to me your line of questioning is completely missing the point with regard to what the environmental sciences are all about---

Examiner Van Susteren: Now that's argument, Dr. Wurster. Can you answer Mr. McLean's question?

Mr. Yannacone: Your Honor, I think [your question] should be just can he answer yes or no [or] does he require an explanation?

Examiner Van Susteren: If he wants to answer yes or no and provide an explanation, then he can do so.

Dr. Wurster: All right, then let me answer this way. I will say no, I do not know much about botflies. I am not an entomologist.

---

*J. Bitman, et al., "Estrogenic activity of o, p' - DDT in the mammalian uterus and avian oviduct," *Science* 162 (1968): 371-372.

phytoplankton

Durable pesticides such as DDT tend to be concentrated when passed up the food chain from plants to small organisms to larger predators; they may reach levels sufficient to wipe out entire species of the animal population.
George Woodwell

I would like to further say . . . why I don't know about botflies. The environment is an extremely complex system. We are only beginning to understand what makes it tick. It consists of a vast array of factors, the appreciation of which requires a great number of specialists. To appreciate this environment, we have got to have botanists and ecologists, zoologists, ichthyologists and ornithologists; we have got to have statisticians and entomologists, agricultural experts, meteorologists———

Examiner Van Susteren: But now, this does not refer to the question.

A: I will get to it, I think.

No one person can concentrate sufficiently in any one field to begin to grasp the complexity of this [system]. And so an environmental scientist must be in the position of being in constant contact with the free flow of information [between] experts and specialists in every conceivable field. Now this I think goes to the heart of the pesticide problem. We are dealing with people in a position to make decisions who are specialists in a narrow field. They may be so simple as to know only that there is a pest and a chemical that will kill it. They are not in contact with the rest of the scientific community.

The environmental scientist is in constant contact with the rest of the scientific community, and my role then is to constantly contact others. If I need to know about a botfly, I am going to have to call up somebody who knows much more about botflies

minnows

mackerel

tern

and entomology than I do. And I'm going to have to evaluate
whether that individual is competent, whether I should accept
his judgment, or whether I should call a different entomologist
after the first one.

I am constantly doing this sort of thing. [So] when you [ask]
me about narrow specialties like this, you are not really getting
to the heart of the issue, I don't feel.

Now let me give you some idea of how the environmental
scientist works. Nobody could consider me an algologist, nobody
in his right mind.

Q: Could consider you a what?

A: An algologist.

Q: Dealing with algae?

A: That's right. Yet, for about the past two years I [have] been rather
suspicious that DDT had something to do with the base of the
oceanic food chains. I hadn't the slightest intention of looking
into it. I just had been wondering what goes on with[in] the
phytoplankton communities in the oceans. And suddenly I found
an opportunity to look into this question.

I spent the summer of 1967 at the Woods Hole Marine Biological
Laboratory. I had access to the best scientists in the world with
regard to algology. They were people who had spent their lives
working with marine phytoplankton. I therefore realized that I was
in a position to begin to look into the phytoplankton question.
I considered this extremely important; it seemed to me that [it]
was something that had to be done, because here you have the
phytoplankton conducting something like 70% of the world's total
photosynthesis. That means that the phytoplankton are responsible for 70% of the carbon fixation on this planet. They are
responsible for 70% of the oxygen in the atmosphere and 70% of
the removal of CO_2. So these little floating algae in the world's
oceans and coastal waters are obviously of exceedingly vital
significance.

I therefore felt it important to look into this question. And yet
here I was without competence, suddenly ... essentially by
chance, . . . in the best place on earth, just about, to conduct some
experiments.

Q: Well---

A: I will get to your botflies.

And so I looked into four different species of phytoplankton.
I investigated diatoms; some little things called *coccolithus* (that's
the genus *coccolithus*); a green alga; and a dinoflagellate. These
were representative of four major groups of algae. They are major
food items for the various marine animals. They are representative
of the base of oceanic food chains.

The results of these experiments were, in a few words, that a
few parts per billion of DDT in the water reduce the photosyn-

thesis [occurring in these algae]. By the time we were up to 10 to 15 parts per billion in the water, the photosynthesis of these algae was down in the neighborhood of 30 to 40% of normal. . . .

Now if I had not been in the Woods Hole Research Laboratory and I had decided I would like to go into that question about the phytoplankton, it would have taken me a year before I could have even gotten those things to grow. I don't have a competency in algology.

But I worked right in the laboratory with Robert Guillard. He's a world authority on phytoplankton. And so the first time that I grew these cultures, they grew, they worked. I was able to employ the services of his technician who has been doing this for years. And she knows how to prepare the medium [on which the algae grow] so it will work. So I was able to conduct experiments in two months that otherwise would have taken me two years.

Now let me read from the back of this paper which is in our technical---

Q: Which paper are you referring to?

A: I am referring to a paper called "DDT reduces photosynthesis by marine phytoplankton." It's from *Science* Volume 159, page 1474.

Q: And what was the date?

A: The date is 29th March 1968.

Now this was written by me, alone, as a single author. But that doesn't really tell the story. The story is told really in the acknowledgement which says, "I think, the Woods Hole Oceanographic Institution---

Q: I'm going to have to interrupt. I'm sorry, Dr. Wurster, I don't like to interrupt a witness. But I'm going to have to make a request at this time that when you, as you refer to studies extemporaneously, I think that Counsel will agree they should be made available so I have a chance to read them.

Examiner Van Susteren: Well that's far afield from the botfly. Could we get back to the original question that Mr. McLean asked in regard to botflies?

5

The Birdwatchers

It is possible for DDT to wipe out an entire species of bird without killing even one individual.   **Dr. Charles Wurster**

In the nineteenth century, miners would bring caged birds into the pits with them. When the birds stopped chirping, the men knew that something was wrong with the air and would return to the surface. Similarly, it was the birds which first told scientists that all was not well with the environment, and tracing the cause of the decline of certain species of raptorial birds such as hawks and eagles led researchers to the first awareness of the sublethal effects of DDT.

Even though DDT was touted as the ultimate insecticide when it first came into use during World War II, the outlines of trouble started to appear early. However, those outlines didn't begin to be shaded in until the mid-1960's when Dr. William Ratcliffe, an English ornithologist, began searching for clues to why the peregrine falcon and golden eagle populations were crashing. He discovered by comparing the weights of eggs found in museums that the shells of eggs produced by birds since the introduction of the chlorinated hydrocarbons were significantly thinner than those produced before the war. Other independent researchers suggested a possible reason for this; DDT was acting as an hepatic (liver) *enzyme* inducer and was thereby interfering with the *metabolic* processes of the birds, especially the process of calcium metabolism crucial to the production of eggs. As the pieces began to come together, a startling picture emerged. The birds atop the food chain slowly were being wiped out by amounts of DDT not large enough to kill them, but large enough to drastically interfere with their reproduction.

Three specialists, testifying for the petitioners at Wisconsin, addressed themselves to this and related problems, each speaking from a different angle to complete the picture of avian catastrophe observed by Ratcliffe and outlined in Wurster's testimony. Either separately or compositely, the three witnesses, Drs. Robert Risebrough, Joseph Hickey, and Lucille Stickel were impossible for the industry to batter down.

The first to appear and most crucial to the anti-DDT presentation was a molecular biologist turned semiprofessional ornithologist, Dr. Risebrough. The enigmatic 34-year-old University of California researcher made a haughty, almost arrogant witness, a perplexing combination of the old and the new in science. But of the originality and perceptiveness of his work, ranging beyond birds to cover in detail many aspects of DDT, there could be little doubt.

Risebrough served an extremely important function for the forces at Madison: as well as presenting much pertinent testimony, he was able to hurdle the Task Force's most carefully laid stumbling blocks, substances called polychlorinated biphenyls (PCBs). PCBs, which are used as plasticizers by a number of companies throughout the world, function as enzyme inducers in a manner similar to DDT and its metabolites. They also have the nasty habit under certain conditions of appearing very like DDT on *gas chromatographs,* causing some experts, even impartial ones, to confuse the two substances in analytic work. Therefore, Risebrough had a two-fold job, to prove that it was primarily DDT and its metabolites, not the PCBs that were the culprits disrupting calcium metabolism in birds, and to prove that he could accurately differentiate between the substances when testing animal tissue samples. Fortunately for Yannacone, shortly before the Madison hearings, Risebrough had developed a method of differentiation and, most important, Risebrough, an expert gas chromatographer—a field more akin to an art than a science—had regularly used that method. Knowing this, the petitioners felt that it was essential to get Risebrough to testify before calling to the stand other experts who would not be as certain of the extremely subtle analytical differences between the ubiquitous plasticizers and the ubiquitous pesticide as he was.

Mr. Yannacone: Now, Dr. Risebrough, have you in the regular course of your professional activities made analyses for the presence of DDT and its metabolites in the environment?

A: Yes, I personally have made over a thousand.

Q: And in the course of these analyses have you had occasion to evaluate the effects and determine the presence of possible interfering substances?

A: For the past year I have been working with substances called polychlorinated biphenyls, and I have now written three papers on this subject.

Q: Are the polychlorinated biphenyls interfering substances in the normal analyses of DDT and its metabolites as residues in tissues?

A: They are not interfering substances in my own analyses, no.

Q: Can they be interfering?

A: In the past some of the polychlorinated biphenyls have interfered with determination of some of the DDT compounds.

Q: Now, Doctor, for the record, would you please tell us just what these polychlorinated biphenyls are?

A: Polychlorinated biphenyls are, like dieldrin [and] DDT, chlorinated hydrocarbons, and their structure is a *biphenyl*. These biphenyls are chlorinated to various degrees. I believe they [PCBs] came into use in the 1930's initially, and then they came into widespread use sometime in the forties and fifties.

Q: What is their common use, if you know?

A: As I understand it, they are used to make plastics, resins, [they are in] paints, various kinds of electrical insulators, and in at least some rubbers. . . .

Q: Now what is it about them that causes them to [interfere] in analyses of DDT?

A: Well, first of all, like DDT they are very stable; [it is] virtually impossible to break them down. And, like DDT, once they are used for their purpose they become waste products. . . .

Q: Now is it possible to mistake polychlorinated biphenyls for one of the DDT metabolites in a gas chromatographic analysis?

A: . . . In analyzing . . . fish, for example, we know that the chlorinated hydrocarbons are in the fat tissues. So the first step in the analysis is to isolate the fat. Then you have a mixture of fat and chlorinated hydrocarbon. The next step is to get rid of the fat. So you end up with a mixture of the various chlorinated hydrocarbons in an organic solvent such as hexane. A known amount of this is put in the gas chromatograph and what comes out is a mixture of peaks.

Q: Would you tell us, please, just what a gas chromatograph does.

A: The gas chromatograph is maintained at a high temperature, something like 200 degrees centigrade. It consists of a solid support of a powder-like material to which a resin is attached. [These materials are contained] in a long U-shaped tube through which a gas is passed. . . . Your mixture of chlorinated hydrocarbons, [placed in] a little syringe, [is introduced] through a septum . . . in[to] the gas stream going through the column. Some of these compounds will come out earlier than others because of their physical-chemical characteristics. The first one that comes out passes through a detector. (The most commonly used detector is an electron capture detector.) It makes a peak in the recorder, so the recorder pin suddenly makes a peak [on] a moving sheet of paper.

It occasionally happens that two of these compounds just might come out at exactly the same time. If so, together they would make one peak.

Then followed an extremely technical description of the subtle differences in the peaking time of the PCBs and DDT in different types of chromatographic columns. Risebrough, who rarely bothered to restrict his technical vocabulary unless prodded by Yannacone, nevertheless made his point clear. An expert could tell the difference between the two substances.

Mr. Yannacone: Is it now common to find DDT metabolites and the polychlorinated biphenyls in residues of tissue?

A: Yes, like DDT compounds, the polychlorinated biphenyls are now found over the world. I have not yet found them in penguins, for example, from Antarctica; I have found only DDE residues. But I have found PCBs, just like the DDT compounds, in sea birds from Alaska, New Zealand, Chile, Australia, and Antarctica. They are worldwide pollutants at this stage. And we believe that, like the DDT compounds, they are dispersed by air transport.

Later, Risebrough added his observations of fish in Lake Michigan to those of birds, saying that they, too, were contaminated with polychlorinated biphenyls, but to a much lesser extent than they were with DDT. This difference in quantity was to prove crucial in later testimony. He continued:

Dr. Risebrough: In general, the [DDT] levels in Lake Michigan fish are higher than those in any other freshwater fish in the country except possibly in some regions of the lower Mississippi. . . . I don't want to say they are almost as high as those in the Pacific Ocean, but it's almost true. They are perhaps twice as high as the residues in fish from certain areas in the coastal waters of California.

The testimony then turned to the levels of DDT in the birds and fish of those coastal waters and, finally, back to the idea that, as far as the mechanisms by which DDT enters living systems were concerned, Lake Michigan could essentially be considered an inland sea.

Dr. Risebrough: In our work we had to account for the very high levels of DDT found in sea birds and marine fish. When we first began our work we assumed that fish in San Francisco Bay, for example (because the Bay drains the San Joaquin and Sacramento valleys, which have very heavy biocide use), would have very high levels [of DDT] and the same species of fish out in the ocean would have no or almost undetectable levels. This is not so. . . .

  Now we also had to account for the fact that the bird with the highest DDT levels . . . in California is not a resident of the cities or of agricultural plants in the valley. It's a species of petrel which never comes in sight of land except to breed on the offshore islands. This is the ashy petrel and the residue levels in this bird averaged about 60 parts per million. . . .

  . . . When we first became aware that levels in . . . birds closely related to the petrel [were] 10, 20, 30, 50 parts per million, we realized that something was wrong in the traditional concept that chlorinated hydrocarbons entered the sea from rivers. We just could not explain these high levels in offshore birds by assuming [the pesticides] were coming from the rivers. . . .

Here in Wisconsin a study has been made of the level in fish in a stream draining an orchard. . . .* The levels . . . recorded in those fish are lower than those in marine fish, and they are much lower than those in Lake Michigan.

In looking at Lake Michigan, you can almost consider it as an inland sea. . . . The interesting thing about Lake Michigan is that the flow between Lake Michigan and Lake Huron goes in both directions and that there is very little water flow in the southern part of the lake, so that input of things like chlorinated hydrocarbons represents much input and very little output. . . .

Mr. Yannacone: In other words, you are referring to the deep ocean basically as a basin and you are referring to Lake Michigan as a similar basin?

A: In which these pesticides accumulate, and once they get there they stay, and that's the only way I think we can explain the high [DDT] levels in fish in Lake Michigan.

Q: Now can you tell us whether or not the accumulation and the transfer or the transportation of DDT and its metabolites is dependent upon any of the physical-chemical properties of DDT and its metabolites? . . .

A: My opinion is—it's not only opinion, it's knowledge—that DDT applied anywhere in the environment will leave that area because of two properties. One of these is its tendency to pass off, to become a gas. If we leave some DDT down here on the table it will gradually become a gas. We may think [it remains] inside, but as soon as somebody opens the door, out into the environment it goes.

The other property . . . is that of co-distillation. . . . If DDT is found on the surface of water, it doesn't like to enter that water, but it will tend to escape into the air, . . . so that when DDT is applied to a [wet] field . . . we can expect it to go into the air even faster than it will by itself. But the end result is the same, it eventually becomes a gas, and when it's a gas, it goes into the global circulation patterns.

I might mention one other dispersal mechanism. When DDT was first used, it was applied with a substance called talcum powder. . . . Several of my co-workers at the Scripps Institution of Oceanography have studied the distribution of [talc] in the global ecosystem, and they find [first,] . . . that the mineral talc can now be detected in the air anywhere in the world by a very sensitive X-ray technique; secondly, that talc can be found in glaciers after 1940, but not before. This means when DDT came into widespread general use in the early forties during the war, much of this applied with talc must have escaped into the general air circulation at

---

*R. J. Mobry and Myrdal Helm, *Pesticides Monitoring Journal* 1 (1968): 27.

that time and deposited in glaciers. . . . This is one way we can conclude that DDT has entered the world circulation patterns.

Yannacone then brought Risebrough over to his second task: that of differentiating between the effectiveness of the DDT and the PCB compounds as enzyme inducers. Though not presented as a continuous story, during the course of examination, cross-examination, and redirect, Risebrough gradually unfolded the evidence.

Dr. Risebrough: . . . [In June 1966] Mr. Walker from Arizona had visited [the] nest of a peregrine falcon . . . when there were three unhatched eggs. Last year we decided to go back to that area at about the same time. And since the peregrine falcon had suddenly become extinct over much of North America and since it was commonly believed that this extinction process might be due to environmental pollutants such as DDT, we were very interested in obtaining one of these eggs, if undeveloped, for chemical analysis. We went to the site. The parent birds had abandoned the nest. There was one unhatched egg, which we brought back to the laboratory, and I proceeded to analyze it.

There was a very high amount of DDE in this egg, five milligrams, approximately 100 parts per million on a wet weight basis.

This egg was collected in an area hundreds of miles from any possible source—let's say a hundred miles from any possible source of DDT contamination.

There were a number of peaks on the gas chromatograph which I could not identify. Since at that time it appeared very important that perhaps these peaks . . . contributed to the extinction of the peregrine falcon, . . . I therefore made a great deal of effort to try to identify these substances, without any success until reports appeared from Sweden that substances called polychlorinated biphenyls had been detected in fish and birds. I therefore obtained samples of these compounds, and after many experiments, concluded that polychlorinated biphenyls were also present in this egg at a ratio of approximately one tenth of the amount of DDE. . . .

We then proceeded to analyze a number of other birds, particularly birds of prey and sea birds for these PCB compounds, . . . to determine the general distribution [of PCBs] in the ecosystem. . . . We . . . wondered if PCBs [like DDT,] induced enzymes, and we drew up an experiment to test this.

In the course of this experiment we decided, since various other papers in the past had been concerned with para, para-prime DDT, which is not the principal pollutant among the DDT compounds in the environment, that we should try to study as much as possible the ecologically relevant compounds, in this case para, para-prime DDE. . . .

We also wanted to use compound[s] . . . which [have] been referred to previously during this hearing; . . . the steroid compounds *progesterone* and *testosterone* . . . which are natural components of animal tissues, to determine whether or not the enzymes induced by . . . DDT and dieldrin, could degrade these *steroids*. However it's much more relevant for my purpose not to use progesterone and testosterone, but [instead, another hormone] *estradiol* [which is found] in birds . . . . because estradiol (estrogen) is intimately involved with calcium metabolism. . . . We thought that if we could show these enzymes broke down estrogen, we could make this more ecologically relevant. . . .

The birds were injected [with DDT, DDE, and PCB] compounds and after several days they were killed. Liver preparations were made and these were incubated with radioactive estradiol. We found that each of these compounds, while in the livers of the birds, had induced hepatic enzymes which could degrade estrogen. In the course of this we found that DDE had the same enzyme-inducing capacity in birds as DDT.

Previously in mammals, in a paper by Hart and Fouts,* . . . it had been reported that DDE has an effect as great or greater than DDT in the induction of several drug-metabolizing enzyme activities when injected or fed to rats. Assuming that we as people have the same enzyme-inducing capacity as rats, we come to the dilemma that even [ocean] fish in a daily diet are sufficient to increase steroid *hydroxylase* activity in people. And fish . . . in Lake Michigan, having higher residues of DDE, would increase the level of steroid hydroxylase activity in the human liver, assuming . . . that human [liver] in this case is equivalent to that of the rat. . . .

Parenthetically, Dr. Risebrough added:

Dr. Risebrough: We find here that the Federal Food and Drug Administration is several years behind research. There is no mechanism in the establishment of tolerance levels to consider this enzyme-inducing phenomenon in people. Rather, they deal with the parts per million approach, how many parts per million it takes to kill a rat, rather than to consider what those compounds might actually be doing.

Then Yannacone steered the testimony, once more, back to the main issue.

Mr. Yannacone: Now, Doctor, this phenomenon of enzyme induction which induces steroid breakdown, is it your testimony that either

---

Archives of Experimental Pathology and Pharmacology 249: 486.

the polychlorinated biphenyls or DDT acting independently can cause such a phenomenon?

A: We concluded that they act independently.

Later, during redirect examination, Risebrough testified that, on a weight basis, polychlorinated biphenyls have a greater enzyme-inducing capacity than DDT or DDE. However, he felt that PCBs might induce enzyme effects by a different route. Risebrough then summarized his findings by saying that the polychlorinated biphenyls and DDE would both cause the thin eggshell phenomenon, but that since DDE was the more abundant compound, he believed that it was principally responsible.

Although the definitive testimony had now been presented, there was still a chance that the processes of enzyme induction would not be understood by the layman. Risebrough embellished.

Dr. Risebrough: I thought it just might clarify the understanding of the general phenomenon, if I gave an interpretation of how it is that we have these enzymes that break down our sex hormones and steroids. Perhaps at some distant time in the past our hypothetical ancestors, men and mammals and birds and fish, would eat a foreign fat or foreign nonpolar substance, pine tar or resin. This still happens, we do eat various kinds of poisons, which are not polar, and once eaten they cannot pass through the kidney, because they are not soluble in water. So at all costs they must be gotten rid of. What appears to happen is that the liver makes a huge amount of enzymes which hydroxylate these compounds, that is, they make them more soluble in water so that they can pass out through the kidney.

This is the primary concern of the organism at that moment: to get the substance out, or it dies. And this seems to have happened in evolution. [Consequently, surviving] fish have these enzymes (DDT can induce [them] in trout); birds certainly have them; mammals certainly have them.

Now, however, the liver doesn't know what is going to be eaten next, what kind of pine tar or wax. It's got to be ready for everything, almost. If it isn't ready, the organism dies. . . .

Now what seems to be happening [is that,] in this need to be ready for everything that comes along, [enzymes] also have the capacity to hydroxylate some of the endogenous steroids at the same time.

During the course of this testimony, Risebrough posed the question, ". . . what happens when these enzymes are synthesized all the time?" The answer, unfortunately, was uncertain.

Dr. Risebrough: In mammals, as I understand it, nobody knows [what will happen]. There is no provision within the Food and Drug

Administration to consider the effect of the constant nibbling away at our steroids. This is not provided for in any of these tolerance things, which are way out of date and which do not even consider PCB, one of the most important environmental contaminants.

Risebrough then returned to birds, relating the material he had just presented to them, specifically.*

Dr. Risebrough: Birds have a problem when they have to lay their eggs. All of a sudden, within a space of a few days they have to

---

*The following material is summarized from Kenneth Simkiss, *Calcium in Reproductive Physiology* (New York: Reinhold, 1967).

My opinion is that there is no possible way in which DDT can be used which will prevent it from entering the ecosystem and thereby becoming both a technological waste product and a pollutant.
Robert Risebrough

provide an immense amount of calcium to make their eggshells. . . . They do this by laying down a special kind of bone
that's called medullary bone; it's found in what is normally the
hollow regions of the bones of the bird. The formation of this
bone is under the control of estrogen in birds.

   . . . As the breeding cycle of the birds develops, as estrogen is
secreted, the birds begin to lay down in what is normally the
hollow part of the bone this medullary bone. One can take male
birds which don't form this medullary bone and inject estrogen
[a female hormone] and these male birds also form this medullary
bone. This bone is the source of the calcium that goes into the
eggshell.

   Now scientists everywhere—Dr. Ratcliffe in Great Britain, Dr.
Hickey of the University of Wisconsin, Dr. Hickey's graduate
student Daniel Anderson—are finding that fish-eating and raptorial birds which accumulate high amounts of chlorinated hydrocarbons are laying thin-shelled eggs.

   We believe that this is a major cause of their decline over the
past years, and we believe that the major reason for this aberrant
calcium physiology is the *in vivo* destruction of estrogen, pre-
venting—this is our interpretation of the data accumulated—the
formation of this kind of bone. . . .

   How was enzyme induction discovered? Yannacone next led the
way to this, introducing the subject with questions related to the
detoxification of certain drugs within the body.

Mr. Yannacone: Is enzyme induction a newly discovered phenomenon, or is it something that's been known for a while?
A: It has been known for several years.
Q: And has it been recognized in certain effects on the relative
toxicity of certain substances such as the phenobarbitals?
A: Well as I understand toxicology, many foreign substances induce
these enzymes. Now these are foreign compounds, and in many
cases they are poisons. The enzymes they induce are detoxifying
enzymes. Things like phenobarbital will be metabolized and
thereby excreted, and therefore the physiological effect will be
decreased.

   Chlorinated hydrocarbons were discovered accidentally in an
experiment with one of the drugs related to phenobarbital. [The
experimental] rats were kept in a room which had been sprayed
with chlordane [a chlorinated hydrocarbon], and . . . experimenters found that these rats . . . were metabolizing or getting rid of
[this drug] or [its] physiological effects were decreased. . . .
Q: In other words, something was helping them generate an enzyme
which broke down the drug?
A: Yes.

Although Risebrough then went on to analyze the evidence of others, showing that DDT affects the reproductive capacity of the coho salmon in Lake Michigan, his direct testimony was, essentially, ended; Yannacone had made his point, and it was up to McLean to refute it.

During the initial cross-examination of Risebrough, McLean tried to sabotage the notion that the supposed disappearance of species of birds, especially the peregrine, was a recent development.

Mr. McLean: It is true, isn't it, that at least a hundred bird species have disappeared in the last few hundred years?

Dr. Risebrough: Yes, that is true.

Q: One of my favorite sources is the 1888 edition of the *Encyclopaedia Britannica*.

Do you know [that], under the heading of Peregrine Falcon in the *Britannicas* of that vintage, the substantial reduction in numbers of the peregrine falcon was a matter of considerable discussion?

A: It may have been a matter of considerable discussion, but it wasn't true . . . even in the species used for falconry throughout the Middle Ages. . . . The numbers of the peregrine falcon in Great Britain remained essentially constant until the Second World War, at which time the War Ministry of Great Britain decided that the peregrines were taking too many pigeons which were used to transmit messages back and forth to France. They issued a decree saying the peregrines shall be exterminated. The numbers of peregrines consequently decreased substantially until the end of the war. Within a very short time, a few years, they were back to their pre-war levels at all the sites in Great Britain. [Then] suddenly the population began to crash. [This] was a sudden phenomenon. Dr. Ratcliffe studied this over the years. He first concluded that perhaps these organochlorine [chlorinated hydrocarbon] compounds were responsible. He noticed some instances of aberrant behavior; he noted many instances of broken eggs. . . . This led him to measure the thickness of the shell of [the peregrine] eggs and the weight of the total eggshells.

Now fortunately or unfortunately, the peregrine has been of special interest to egg collectors, so that both in Great Britain and this country there are many shelves full of peregrine eggs. Dr. Ratcliffe went to those shelves; he measured the thickness of the shell and he measured the weight of the shell. He found that up until about 1945 the weight of the shells had remained constant over the many years. After—I believe, 1947—he noticed a decline, a dramatic decline. And peregrines have laid thin eggshells ever since in some regions. . . .

McLean made a few parries and thrusts at Risebrough about a number of subjects, such as how this or that species of bird was faring—pointing to the fact that gallinaceous birds such as pheasants did not show the thin eggshell syndrome, and receiving a reason from Risebrough for this—and how population stress influenced eggshell thickness. Then he went after Risebrough's credentials, eliciting a poetical response that did little to aid the DDT industry's cause.

Mr. McLean: Doctor, you stated that your educational background is that of a molecular biologist; and I believe you also said you have made in excess of a thousand analyses by gas chromatograph?

Dr. Risebrough: My thesis was in molecular biology.

Q: Which do you consider yourself now?

A: An environmental scientist.

Q: Does that mean that you do not classify yourself now as a molecular biologist or analytic chemist?

A: I do not believe in pigeonholing people. I consider myself, if I have to have a label, as an environmental scientist. And I think it's precisely because people have considered themselves specialists that very few people realize what's going on in the environment: for example, how pollutants and the amount of waste products produced by man in the course of a year are now equivalent to the amount of carbon fixed by plants. This means that the whole framework in which science has worked up until now has changed. Our environment is no longer what it was. It's in great danger. And to tackle these problems, one does not go in and measure the peaks that come off the gas chromatograph; that does not tell you what's going to happen in Lake Michigan.

Risebrough was home clean through the two days of cross-examination by McLean, who couldn't perturb the cool, bearded environmentalist. However, when McLean was replaced by Stafford after the long recess, Risebrough was almost in trouble. Stafford read from an article in the *Los Angeles Times* by Irving Benglesdorf and asked if Risebrough were quoted correctly.

Risebrough said that the quotation was another scientist's interpretation of some of his findings; Risebrough's view, stated correctly, was that the polychlorinated biphenyls on a weight basis have about five times the estradiol degradation potential of DDE.

Stafford then read from the newspaper article: "peregrine falcons, which contain extremely high concentrations of PCBs, up to 98 ppm., now are extinct in the eastern United States" and asked whether this was a fact.

Risebrough replied:

Dr. Risebrough: The information imparted is incorrect. I have measured a PCB concentration of 98 ppm in the peregrine falcon that was found dead in the San Francisco Bay area. That particular bird, if I remember correctly, contained a still higher concentration of DDE. We can't extrapolate from the PCB concentrations in that bird to the peregrines in the eastern U.S. . . .

There is abundant evidence now that the peregrine falcons in the eastern United States lay thin eggshells and that this phenomenon was associated with their decline. All available evidence . . . that we have now suggests that DDE is much more responsible for the thin eggshells than is PCB. . . . At the moment, we have no proof that PCB causes thin eggshell formation. . . .

One of my areas of research has been local fallout patterns. One of the clues to fallout is the ratio of DDT compounds to PCBs, and they seem to be characteristic of a given locality. . . . I have concluded that there are global aerial fallout patterns of these compounds. . . . In the Pacific, the ratio of DDT to PCB is something on the order of 10 . . . times as much DDT as PCB. . . .

Van Susteren then asked whether the thin eggshell effect could be traced specifically to DDT or PCB when both were found in a single specimen. Risebrough said that this had been done.

Dr. Risebrough: [We] have concluded that, in the case of the white pelicans, DDE is causative of thin eggshells and PCB is not. . . . This was determined by collecting a large number of eggs of white pelicans. . . . These eggs were subsequently analyzed for both DDE and PCB. The concentration of both was very low. The highest concentration of DDE was no more than 6 ppm in the egg but significant shell thinning was observed. Correlation of shell thickness and DDE content of the egg was significant at the 95% level. There was no significant effect of the PCB content of the egg upon shell thickness. . . .

Referring to an unpublished paper he had prepared on analyses to separate PCBs and DDT compounds, Risebrough said:

Dr. Risebrough: In the petrels and shearwaters, PCB is perhaps one tenth the amount of the DDT in those pelagic species from the open Pacific. There are several species of petrels and shearwaters from off the California coast and I think they might have had about five times as much DDT as PCB. I said that there the concentrations of PCB are high. That's true, but the concentration of DDE was as high to 10 times higher still. . . . I have yet to measure any bird sample, including eggs, which had considerably more PCB than the DDT compounds.

Dr. Risebrough was off the hook; the biochemistry was clear. It was now up to Dr. Hickey and Dr. Stickel after the hearing recess to complete the story.

6

The Recess:
the Laboratory, the Streets
and the Government

Spokesmen for the DDT industry often talked in derisive terms of the "bird watcher" scientists who comprised the bulk of the petitioner's forces, and of the distance between those lofty academic personages and the realities of the world. But that world's realities burst upon the scene in a number of ways during the Madison hearings.

The University of Wisconsin has not built up a reputation as the most placid campus in the nation and has often vied with Berkeley as the place most likely at any given moment to have a riot—and those given moments occurred again and again. During the first week of the hearings, when proceedings were still being held in the State Capitol Building, the Hearing Examiner made formal acknowledgment of a "commotion" going on outside the hearing room. This innocuous acknowledgment gave no indication that the commotion was being made by hundreds of angry students trying to force their way into a meeting of the State Board of Regents to protest the expulsion of 100 black students from Oshkosh State College.

But that demonstration was just a warm-up. In February the campus erupted with a strike over the demands of campus blacks which featured, before it ran its course, the calling in of the National Guard to protect all matter of living and non-living things except striking students, and the gentle bathing of the town in tear gas. Then in April, while arguments both subtle and direct were taking place inside the DDT hearing room, firebombs were being tossed and heads cracked in what was at the time the biggest single uprising in Madison's history.

Attitudes toward these commotions and those causing them varied greatly within the anti-DDT coalition, running the gamut from those petitioners who felt basically sympathetic with the students to those who felt basic hatred toward them. (Some of the members of the coalition were people who felt politically and socially more at ease with the gentlemen of industry than with the amalgam of beards and youth.) Yet, one of the more significant developments in what promises to become the student ecological battle of the 1970's surfaced as the result of those Madison hearings. It was there that the first

71

ecological activist tactics of the current fad took place, with the formation of the DDT Commandos. These Commandos were the action arm of the Conservation Research and Action Project (CRAP), in turn a spin-off of the University Science Student Union, a group of some Wisconsin radical science undergraduate and graduate students concerned about the use and misuse of science in society.

The hearings, with their national publicity, naturally attracted the interest of the Science Union, and almost every day bearded, bedraggled radical students could be seen avidly watching the goings on. From their interest in the hearings came CRAP, and CRAP's first move was to sic the Commandos on the proceedings.

The group's actions on that winter day were pure guerrilla theatre. About 25 students, dressed in makeshift commando costumes, and armed with placards emblazoned with such revolutionary slogans as "Liberate the Ecosystem," "Get the Crap out of the Environment," and "Ban the Bug Bomb," invaded the decorous State Capitol Building after marching on or, rather, stealthily attacking it from a nearby area of the campus. These Commandos carried water guns, claimed to be filled with DDT, that they proceeded to squirt at everything in sight. So, in best Western melodrama style, a Wisconsin Marshal disarmed them at the door of the hearing room, and forced the ferocious fighters to stack their water weapons. Of course, while much of the student population of Madison flipped at the Commandos' actions and the media men outdid each other taking pictures, the coalition sternly disapproved. Even Yannacone drew the line and, feeling that the Commandos were jeopardizing the proceedings, requested that they wage war in other areas. Industry, too, took the Commandos and their cohorts seriously, going to the length of editorializing in one of their trade publications that the Commandos and the scientists testifying for the anti-DDT coalition were equatable. Nevertheless, the Commandos planned further action for the spring, but the vicious street fighting which broke out in Madison at that time took away the initiative to have such fun.

Shortly after the winter "DDT offensive" which successfully brought the Commandos to the doors of the hearing room, the hearings were moved to another location some three miles from the University.

But, while student antics were providing comic relief to the proceedings, another event was occurring in neighboring Michigan which directly affected widespread public feeling about DDT.

Predictions of this event were heard in the 1968 trial before the Michigan Court of Appeals when Dr. Wurster prophetically testified that the levels of DDT and dieldrin being found in the Lake Michigan ecosystem would be enough to affect the reproduction of its coho salmon. In the most ambitious fish transplanting endeavor ever accomplished, these West Coast salmon were introduced into Lake Michigan to replace the lake trout which had been almost annihilated by the sea lamprey. At first, the salmon thrived almost beyond the wildest dream of the wildest fishery biologist, and soon Michigan was planning on a tourist-attracting industry that would increase the state's coffers by millions of dollars every year. But Wurster was proven frighteningly right; the death of a million coho fry the next spring was attributed to DDT.* As could be expected, this caused quite a stir among conservationist anglers. But the full implications of these DDT levels were not apparent even then; things were to get worse.

In February 1969, the pesticide level in a can of coho salmon was routinely checked by the Consumer Protection Division of the Michigan Department of Agriculture. It was found to have a dieldrin level of .32 parts per million. At that time, there was no official Federal Food and Drug Administration tolerance level for DDT in fish; however, the FDA had established a guideline level in fish of 0.3 ppm for dieldrin, a chlorinated hydrocarbon similar in environmental characteristics to DDT. A "Stop Sale" was immediately placed on 146 cases of salmon from that specific lot, and the Consumer Protection Division investigated further lots for high dieldrin level. According to Dr. George Whitehead who headed the department, they found no other cans which exceeded .3 ppm.

What the doctor neglected to mention, however, was the positively astounding levels of DDT, ranging up to 50 ppm, in those cans; those figures were stamped "confidential," and the public was not to learn about them until the affair exploded three months later during the Madison recess. Dr. Whitehead claimed afterward that he did nothing about the excessively high DDT levels because the federal government, whose guidelines he accepted on dieldrin, had not released tolerance figures for DDT. But, as some critics harshly said after the affair came into the open, with the FDA tolerance level for DDT in beef at 7 ppm, what was Whitehead waiting for?

Tom Brown, director of the FDA regional office in Detroit, unlike Whitehead, did not wait for an official tolerance level to be established. Despite the absence of a guideline level, he moved in and banned the sale of 32,000 pounds of coho salmon on the grounds of potential danger to human health. He claimed that he didn't wish to wait three years and then find out that the fish were dangerous. As could be expected, the uproar was fantastic. The public health agencies said that there was no proof that the fish presented a danger to human health; the agricultural department in Michigan coyly said that there was no real reason to ban DDT; the public and the Michigan Department of Natural Resources under Ralph MacMullan, a persistent foe of persistent pesticides, bellowed that the stuff must go.

As a result of the outcry, two weeks after the FDA banned the sale of the salmon, the Michigan Department of Agriculture's executive board met and banned DDT.

---

*DDT kills the fry after they have hatched but before they fully resorb their yolk sacs.

*Perhaps, as with every disaster, natural or man-made, no
one believed it would happen.*
N. Y. Times *August 9, 1970*

Even before Michigan knocked DDT off its shelves, some 6,000
miles away, the National Poisons and Pesticides Board of Sweden
decided that the Scandanavian environment had all of the chlorinated
hydrocarbons that it should have and announced a ban on DDT use
for two years, a ban that everyone in the country expected would
last indefinitely. The Board also decided to ban the use of diedrin
and aldrin in Sweden as well.

At a conference held by the board which terminated in DDT's
demise, a Swedish scientist, Dr. Oden, estimated that the total amount
of DDT and its metabolites in Sweden's soil was 2,000 metric tons—
probably more than had ever been used in that country. But perhaps
the most important testimony at the Swedish hearing came from a
man who was to figure prominently at Madison, Dr. Goran Lofroth.
This young scientist testified that DDT from foods and from the
environment was pervading the Swedish biosphere to such an extent
that it was appearing in sizable amounts in human milk. In a country
where breast feeding is common, this aroused the ire of many a
household.

Dr. Wayland Hayes, a representative of the DDT industry and a long
time defender of the human non-toxicity of DDT, appeared in Sweden
to refute the quickly accumulating evidence. Hayes's data had long
been a stumbling block for the foes of DDT. In the past, scientific
opposition to the chemical had often dissipated in the face of his

evidence based on experiments which he conducted in the United States during the 1950's on prisoners and workers in pesticide factories. But at the Swedish hearing Hayes and Lofroth cracked heads, and Lofroth emerged as the winner by a knockout.

So, with a conclusive win already on the records, Lofroth was booked for a second scientific confrontation with Hayes at Madison. And again, although the result was less conclusive, most outsiders agreed that Lofroth, by showing the degree that DDT has pervaded human life and by casting grave doubts on Hayes's cant of the innocuousness of the pesticide, came out on top.

Of course with any sort of controversy raging, California was sure to jump in with flourishes of its own. All sorts of bills regulating and banning the use of chlorinated hydrocarbons were bandied around the state legislature. Heated arguments erupted pitting various departments of the various state universities against each other. (As an interesting sidelight here, Ray Smith, head of one of the most prestigious departments of entomology in the world, the one at Berkeley, came out against a bill banning DDT while his most prominent pupil, Robert van den Bosch, was testifying in behalf of the petitioners in Wisconsin.*)

The long-raging grape pickers, schism causers for several years in the California social scene, found an issue within an issue with DDT. Led by Caesar Chavez, the grape strikers claimed that DDT along with other pesticides was poisoning not only them but the grapes themselves. The grape pickers then made an attempt to get information on pesticide use from the County Agricultural Commission in Delano, headquarters of the strike, only to learn the information was considered by California to be "proprietary" and thereby more important in itself than were the lives of the workers.

Finally, a bill was passed in California which banned the use of DDT in dusting compounds and for household use. But this had little effect since something like 90 per cent of the DDT used in the state is applied in oil solutions.

Jumping on the anti-DDT bandwagon became a popular pursuit in a great many states during the recess. Maine restricted much of the pesticide's use as did Pennsylvania; Minnesota no longer allowed it to be used for Dutch elm disease control. Almost every state had some sort of pesticide control bill in the hopper before most were killed in agricultural committee. Illinois for instance passed a bill in its lower house 345–0 but the chairman of the Senate Agricultural Committee there, a man who admitted he didn't know awfully much about the chemical, allowed the bill to die in committee. Even Wisconsin did its thing by banning DDT for Dutch elm disease, thereby cutting its use by 50 per cent in the state.

But the hearings went on.

---

*See testimony, pages 115–126.

7

More Birdwatchers

Doctors Joseph Hickey and Lucille Stickel fit the image of the bird-watching scientist. Competent, restrained, positive of the accuracy of their work, the two pinpointed the destruction of raptorial birds outlined by Risebrough, with Hickey supplying the field data and Mrs. Stickel corroborating it with the laboratory work she had done at the Patuxent Wildlife Research Station in Maryland. Neither was a "public figure" scientist like Risebrough and Wurster. Neither was as cocky or as willing to stray into matters philosophical and moral. But they both were crucial to the petitioner's case; and what they said, they knew, and what they knew, they knew for sure.

Hickey, because of a bad cold and a fragmented schedule, was called to the witness stand several times for direct examination, cross-examination, and redirect. But the internationally renowned ecologist remained almost unshakable through it all. Although he seemed to present a lovely target for McLean, in that his subject, the disappearance of birds, was a scientifically fragile matter, a devastating cross-examination never occurred. It never occurred because Hickey never allowed himself to be caught.

His direct testimony was just that, direct and to the point.

Dr. Hickey: Now the thing that I'm particularly interested in as an ecologist is not the mere fact that one or more of these [DDT] compounds is causing the death of a bird. The important thing to me is what is the effect of one or more of these [pesticides] on a population. . . . Because it's easy enough to state that a bird has died of DDT poisoning under certain conditions. But what is the actual effect on the population of the species in some region as large as a county, a city, or half of the United States?

To illustrate his point Hickey focused on the peregrine falcon, a bird he knows very well.

Dr. Hickey: Now to get at these population effects, one faces certain hazards. It is not easy to find birds that die in the wild. But there are some species which can reasonably be censused over fairly large regions. These birds, which are conspicuous nesters, include species like the bald eagle, the osprey, and a species called the peregrine falcon. . . .

Now to get at the changes that have taken place, I need to report for the Examiner something of the population dynamics of these birds. . . . [The peregrine falcon] is a bird that was used in medieval falconry. It is present on nearly a worldwide basis. It nests on cliffs, in some countries on trees, in a few others on bogs and in certain instances on buildings.

The thing that is of particular interest to ecologists is the remarkable population stability of this species. . . . For instance, there is an island off the coast of Wales where peregrines were nesting in the year 1243. In the late 1930's, I discussed this site with a British falconer. The birds were still present. . . . What we have found is that of 49 peregrine falcon eyries (nesting places) known to falconers from the sixteenth century to the nineteenth, 42 were still in use in the 1930's.

Hickey then began describing the decline of the peregrine as occurring in both Europe and the United States at the same time.

Dr. Hickey: When we began to get our first rumors of something wrong with [the peregrine] population on two continents, I'm afraid that it was sort of impossible for scientists to come to grips with a phenomenon that was bound to prove to be so spectacular.

Hickey went on to say that, by 1953, reports of peregrine reproductive failure were appearing in studies on the Hudson River and in Massachusetts. Within the next ten years, similar reports showed up from Britain, Germany, Finland, East Germany, and Northern Ireland; by 1964, it was apparent that the peregrine falcon, as a breeding bird, had been wiped out of the eastern third of the United States.

Then Hickey went on to more details about the decline of other raptorial birds in areas around the world, bringing in a mass of supporting evidence concerning the decline of the bald eagle and the osprey in this country and the golden eagle in Scotland.

As Hickey's story of the effects of the chlorinated hydrocarbons on birds unravelled, it became apparent that it was a tale matching the best detective story—except that the good guys have yet to win.

Dr. Hickey: The peculiar fact that peregrine falcon populations were apparently declining on both continents and that these [declines] were being echoed by other raptorial birds, led in 1965 to the convening of an international conference on the population biology of the peregrine falcon.

I'd like to point out that this conference was not on the effect of insecticides·on the peregrine falcon; it was an attempt to look at the whole picture of the population dynamics of a species common to two continents, and to review population changes in related raptorial birds.

We had representatives of the scientific community from Finland, Germany, Britain, France, Canada, and the United States. . . .

Now what developed at this conference was very interesting. For one thing, we found that the population declines that we were talking about—where we could get data—were characterized by the same symptom: reproductive failure. More importantly, however, was the extraordinary condition of eggshell breakage. . . .

Hickey summarized the reports of this eggshell breakage and its location, then continued to unravel the story.

Dr. Hickey: Now in 1965 when we tried to consider this peculiar condition of eggs being broken and eaten in Britain and eggshells flaking in Wisconsin, the British hypothesis was that we were dealing with central nervous system poisons in the environment and that there were changes in behavior to be expected.

I don't think that the American investigators at this conference were ready to accept a behavioral change as being involved. It meant too complicated an hypothesis. We were convinced, all of us at this meeting, that eggshell breakage had largely been overlooked in North America, that American investigators . . . studying bald eagles, peregrine falcons, and ospreys were not climbing to nests as assiduously as our British colleagues were.

So we were left with the need to find some new explanation for an absolutely extraordinary change in the physiology of these birds, which was certainly present in Britain and apparently taking place also in our own country.

Now the hypothesis began to emerge, derived in part from the work Dr. Welch reported on yesterday. We had this accidental finding that chlordane applied originally to control bedbugs in a laboratory produces– – –

McLean objected at this point, but the information about one of the amazing accidents that make up science still managed, if only in a most fragmentary way, to get into the record. Like Risebrough, Hickey began to tell of the accident which took place in a laboratory at North Dakota State College where experimenters were working on the effects that starvation had on the metabolism of barbituates in rats. Normally, starvation intensifies the effects of such drugs. But in this particular experiment, the reverse took place; the drugs were metabolized more quicky than normal. The scientists conducting the experiments were perplexed. They began asking around the laboratory and discovered that one of the technicians had found that the lab was infested with bedbugs and had sprayed with chlordane, a chlorinated hydrocarbon generically related to DDT. They hypothesized that the pesticide had something to do with the metabolism of the barbituates, and later confirmed this theory.

This accident led to a myriad of studies, many of which were presented at Madison, showing that the chlorinated hydrocarbons induced enzymes in the liver which could break down many substances including steroids.

Hickey continued.

Dr. Hickey: The pharmacological data that followed [constituted] something of a cascade of experimental data on the effect of these chlorinated hydrocarbons on laboratory animals with respect to drug metabolism, the breaking down of steroids in animal bodies, and so on.

Now at this point the ecologist came into the picture, because here we were dealing with some kind of calcium problem. It seemed possible that we were not actually dealing with a change of behavior, as the British suspected, but simply with a physical

> *What is important about the peregrine, aside from its loss—which is a disaster—is the fact that it is the first chemically killed American bird for which documentation of how and why it died has been provided.*
> N.Y. Times *August 9, 1970*

change in the physiology of the birds which involved thin egg-shells.

This roadblock, so to speak, was broken by Ratcliffe in 1967* when he reported, by going to eggshell collections in British museums and private [homes], that an extraordinary change in the thickness and weight of eggshells of certain species had indeed taken place. The species which were reproductively failing, like the peregrine falcon, the golden eagle, and the sparrow hawk, had indeed suffered this change. The other species which Ratcliffe examined that were not exhibiting reproductive failures had not suffered this change.

This, substantiated in minute detail, was Hickey's testimony. However, when Hickey tried to describe the particular physiological factors involved in the decline, the Task Force lawyers objected strenuously, stating that he had no expertise in this area. Hickey was made to stay in the position of a birdwatcher, pure and simple, and it remained for a soft spoken no-nonsense woman, Lucille Stickel, to provide the hard laboratory data necessary to substantiate Hickey's field work. As she did so, the pieces of the birdwatchers' puzzle fit together perfectly.

Mr. Yannacone: Now, Dr. Stickel, will you briefly summarize your work with DDT and its metabolites? . . .

A: The Patuxent Wildlife Research Center conducts studies to determine the effects of environmental pollutants on wild animals and the environment.

Today I will report to you the results of two sets of studies concerning the effects of the organochlorines upon birds. The two studies involve distantly related birds of two important groups: waterfowl and birds of prey. You have heard from others of the serious plight of the peregrines and eagles that are members of the second group. Some members of the first group are declining, also, particularly the black duck, first cousin to the mallard, which I will discuss today. The birds of prey I will discuss are the kestrels, which are close relatives of the peregrines.

Now these studies are experiments, which means that the performance of the birds receiving the organochlorines in their diet is compared with the performance of birds of the same age and history kept under the same conditions but fed untreated food. The birds of the second group are called controls, a term I will use hereafter.

The experiments with mallards resulted in a strictly clear picture. Counts were made of the number of eggs that were cracked. These

---

*D. A. Ratcliffe, "Decrease in eggshell weight in certain birds of prey," *Nature* 215 (1967): 208–210.

eggs were discarded, and sets of sound eggs were chosen at random for measurement of shell thickness.

The remaining eggs were incubated. Those that were infertile or did not develop were discarded. Those that began to develop were continued in the incubator. These were used as the baseline to measure the production of healthy ducklings that hatched and survived to two weeks of age. The hatched ducklings were fed clean food regardless of the diet of their parents.

Now you can see that every conservative precaution was used to insure measurements that were thoroughly objective. Thus, we have measurements of crackage, shell thickness, and success of normal developing eggs.

Mallards were fed diets containing DDT or its metabolites DDD or DDE, each separately. Controls were fed clean food.

DDE was fed [either] at 10 parts per million or 40 parts per million on a dry weight basis. . . .

The results were as follows: ducks fed DDE cracked or broke 24 per cent of their eggs, controls cracked only 4 per cent. . . . Eggs that were not cracked that were produced by the ducks fed DDE had shells 13½ [times] thinner than the controls. . . .

Incubated eggs laid by the ducks fed DDE produced half or less than half as many healthy ducklings as did controls.

Both amounts of DDE produced effects of similar magnitude. In other words, a "no effect" level was not found and must be lower than the dosages fed. . . .

Now I will go to the other compound. This, so far, has related to DDE alone.

DDT, purified para, para-prime, not including the ortho-para fraction, fed at 25 parts per million (roughly equivalent to eight parts per million on a wet weight basis), shows results similar to those produced by DDE at 10 or 40. . . .

It is evident that small amounts of DDE in the diet of mallards affect their reproduction adversely in two important ways: first, eggs are cracked and broken; second, even eggs that begin development fail to produce a normal number of healthy ducklings. It is also evident that thickness of eggshell is associated with these events, and is a strong indicator of serious problems.

The experiments with kestrels, commonly called sparrow hawks, belonging to the genus *Falco*, are equally clear. These, too, involved both controls and dose birds. They were unique in that no predatory bird has previously been reared in captivity with enough success to permit experimental study. A prototype or guinea pig, if you will, was needed for predatory birds. With this in mind, the development of the colony was begun.

By 1967 problems of care had been solved. In 1967 and 1968 eggshell comparisons were made between birds that received clean food and those receiving food containing mixtures of DDT and dieldrin.

I think our robins each spring are newcomers, one year old, or in their second year, because the previous campus populations have been practically exterminated by the first of June each year.
George Wallace

On a wet weight basis the dosages were [either] two parts per million of DDT plus one-third part per million of dieldrin [or] five parts per million of DDT plus one part per million dieldrin together. . . .

Shells of eggs laid by dosed birds were approximately 15 per cent thinner than those of controls. Shells of eggs of the second generation fed the diet of their parents were also approximately 15 per cent thinner than those of their controls.

These results speak for themselves and are in harmony with results with mallards.

From these studies it is evident that DDE in minute amounts can cause marked impairment of the reproductive success among two major bird groups.

McLean's cross-examination of Dr. Stickel provoked a unique response. Dr. Stickel took McLean's questions to impugn her virtue and veracity and, therefore, replied indignantly and chastisingly to the mere attorney who would question her work. Yet, each of McLean's questions was answered deftly and with a businesslike attitude and an amazing—almost intimidating—exactitude.

Mr. McLean: Doctor, for the mallards, I would like to know what the results were in each of the categories that you mentioned: the number of cracked eggs, good eggs, and hatchability. I would like to know that on a pen basis at each feeding level, as well as for the controls.

A: I will be glad to provide the statistical expression of that variability, which you can have interpreted. I cannot provide the raw data sheets that are very long and very thick. But I will provide you adequate data to be verified by any statistician that you can consult. It will express the variability in the manner that you wish and will be valid data. . . . If you wish to visit Patuxent, I will be happy to show you the data sheets.

Then Dr. Stickel pronounced solemnly:

Dr. Stickel: The data is statistically significant. I am somewhat affronted that our veracity is questioned to the point of being requested to see our raw data. But we have nothing to be ashamed of!

McLean sheepishly added:

Mr. McLean: Doctor, just for the record, I hope that you don't think that because lawyers inquire, that we are casting aspersions. This isn't the intent. This is the way we lawyers are.

The final birdwatcher, Dr. George Wallace, was probably the one for whom the hearings at Madison personally meant the most. His scientific testimony was no great shakes but, since the death of Rachel Carson, he could probably be considered the most persistent and long-standing scientific foe of DDT in the nation.

Compared to the new breed of scientist, exemplified by Wurster, Risebrough, Welch, and Steinbach, Wallace was just a naturalist. But, nonetheless, while many of the testifying scientists were in grade school or else hiding behind signs reading, "insufficient data," he was claiming that DDT was responsible not only for mammoth killoffs of robins on the East Lansing campus, but that it was threatening disaster on a much larger scale. Wallace, a professor of zoology, had been speaking up since the 1950's about DDT but had not had available the mammoth outpouring of data presented at Madison.

For the others testifying for the petitioners, the Madison hearings were a crusade against what they saw as a real danger to the planet, but to Wallace they were even more. They were the vindication of two prophets, himself and Rachel Carson.

Wallace spoke of his work with birds on the campus since his arrival there in 1942. He told about how, since 1954, he had watched the Dutch elm program involving DDT on the campus, and how, by 1957, he had watched the robins almost disappear.

He told of robins dying of tremors after the spraying program; he told of the characteristics of DDT-poisoned robins. This was the kind of data that was later substantiated by a scientist of the new breed, Dr. Allen Steinbach, who told of DDT's effects on the nervous system.*

It was singularly appropriate for Yannacone to call Wallace as a witness, for without him the hearings would have had no sense of tradition. It would have seemed to outsiders that they were the invention of the new breed of scientist, when, in actuality, without Wallace, there probably would never have been a *Silent Spring* and never a Madison. It was a good day for the petitioners.

---

*See pages 108–113.

8

The Human Watchers

Although the pesticide industry had, in one form or another, eleven witnesses speaking in behalf of DDT, their case really rested with two men: Harry Hays, head of the U.S. Department of Agriculture's Pesticide Registration Division, who was called to show how well the public was protected from the harmful effects of all pesticides;* and Wayland Hayes, who was there to assert that pesticide protection for man—at least in the case of DDT—wasn't really necessary, anyway. If either of these men could have gotten off the Madison witness stand with their testimony on the safety of DDT to man relatively unscathed, there would have been hope for the future of DDT, no matter what material the petitioners massed on DDT's effects in nonhuman species. For, to the majority of the public, concerned with its human problems, damage to birds and fish and plants would have seemed relatively insignificant. However, these two men did not get off unscathed.

Wayland Hayes, billed by Task Force attorney Willard Stafford as the "world's top toxicologist," had long been a major thorn in the side of the critics of DDT. In the past, whenever the debate over DDT's effects had become too hot and heavy, this dignified southern Public Health Service physician would reassuringly appear and, with best bedside manner, would dutifully inform the public that he himself had conducted extensive tests with DDT on humans, and the stuff, at the levels at which it was being found in the environment, simply wasn't dangerous.

Since his data was really all that had been available in this particular area, ecologists and other hard-pesticide critics had always found combat very difficult. Too often, DDT's critics were made to look like callous scientists, willing to sacrifice the well-being of the hungry masses of the world to the survival of a few birds.

It was essential then, that the Madison petitioners, unlike their predecessors, successfully discredit Hayes's testimony and to do so they were forced into a two-part battle. To dispel the lingering image

---

*See Chapter 10.

of callousness, the petitioners had to prove that DDT was not essential to crop success and disease control; to rally public support, the petitioners had to prove that DDT could, indeed, be considered a hazard to humans as well as to other species. These battles, contested hotly by both the DDT industry and the environmentalists, were what raged so fiercely in and around the testimony of Wayland Hayes.

For Yannacone, planning crucial cross-examination tactics at a time when the effects of DDT on humans were really little more than hypotheses in the minds of a few far-seeing scientists, campaign strategy was frightfully difficult. Primarily, it consisted of deluging the hearing room with enough inferential data on possible human health hazards and establishing enough doubt as to the thoroughness of Hayes's research, to insure the fact that Hayes's findings would never again be held in the esteem they had once been.

Wayland Hayes had received his Ph.D. from the University of Wisconsin in 1942 and his M.D. from the University of Virginia in 1946. He was then employed by the U.S. Public Health Service from 1947 to 1968, when he became a professor of biochemistry at Vanderbilt University.

Under direct examination by Stafford, Hayes listed a number of the papers he had authored and co-authored, and others which corroborated his findings on DDT and humans: "The effect of known repeated oral doses of DDT in man"; "Storage of DDT and DDE in people with different degrees of exposure to DDT"; "Storage and excretion of DDT in starved rats"; "Storage of insecticides in French people"; "Poisoning by DDT: relation between clinical signs and concentration in rat brains"; "DDT storage in the U.S. population"; "Chlorinated insecticides in the body fat of people in India"; "Mortality from pesticides in 1961"; "Monitoring food and people for pesticide content"; "Toxicity of pesticides to man"; "Combined effect of DDT, pyrethrum and piperonyl butoxide on rat liver"; "The pesticide content of human fat tissue"; "Coordination of activities relating to the use of pesticides." These papers made up the bulk of the Gospel according to Hayes.

Stafford then began delving into the substance of the work Hayes had done himself or had monitored, starting off by asking of him a conclusion:

Mr. Stafford: Have you an opinion, based upon a reasonable scientific certainty, whether DDT as now used in this country presents a health hazard to the people?
Dr. Hayes: Yes, sir, I have.
Q: What is your opinion?
A: I think it's safe.
Q: And will you state your reasons, please, in some detail?
A: Well, DDT is used in agriculture, it's used as a pesticide in the home, and in various other ways. No matter how people are

exposed, no matter how it is used, insofar as they are really exposed to the compound, they will absorb some fraction of this material. Therefore, one can evaluate the total exposure in terms of storage of the material measured chemically in samples from people. (I might add [DDT] was used for the control of malaria in this country with great success; and malaria is now eradicated in this country; and so it is no longer used for that purpose here, but it's important for that use in other countries.)

Now going to storage: We have measured the storage of DDT in people in the general population, in people with unusual dietary habits, in volunteers who were given known doses of DDT, and in workers who worked in the manufacture and formulation of this compound and therefore had exposure of a very intense kind for many years.

Then Hayes went into the technicalities of the dosages given volunteers and summarized by saying:

Dr. Hayes: The conclusion was that we could find no effect on the men clinically [caused] by DDT. They complained of no effect that could be traced to this compound. And we examined them [medically] and did a number of laboratory studies on them.

Hayes stated that the dosages fed the volunteers who were, in some experiments, prisoners in sourthern penal institutions, were approximately 200 times what the general population was receiving environmentally at the time. He cited workers in DDT plants who had worked with the compound for over six years and other workers who had been exposed to high levels of the chemical for as long as 19 years, saying that his neurological examinations were "unable to detect any effect on their health." He continued:

Dr. Hayes: As I have said, we could find no effect either by our examination or by examination of their work record. And work record is a rather sensitive measurement of people's physical ability. You find that when they are sick they don't go to work very well.

Mr. Stafford: Now there's been some suggestion in this proceeding—at least it involves certain raptor hawks—that there may be some reproduction problems associated with DDT. Did you find any reproduction problems amongst this group of workers or volunteers?

A: . . . We did assure ourselves that these men in the course of 19 years, for the majority of them, had in fact had families of quite adequate size, a bit above the average as I understand it. . . .

Q: Are there any other studies which you have not referred to which substantiate your conclusion that DDT does not affect health in people?

I think that for people in the general population, [DDT] is completely safe. . . .
Wayland Hayes

A: [DDT] has been used, of course, in disease control in other countries in a variety of ways. The three major diseases that have, in fact, been controlled by it are malaria, plague, and typhus. Its use in these diseases is quite different.

In malaria control, to a very large degree, it is used as an indoor house spray. This involves really little or no increase in exposure of people.

Again, in the control of plague, it is used in the environment and there is very little personal exposure except in workers who use the material.

But in the control of typhus, it's necessary to dust the people. The first time that an epidemic of typhus was controlled or stopped in all history was in Italy during the latter part of World War II. People were dusted with power dusters; thousands of them had DDT dust blown into their clothes, through the collars and the cuffs and down their pants, so that they were pretty well saturated with the powder. . . . The experience with this was that it not only, in fact, did stop this very epidemic disease, but that there were no ill effects from its use. . . .

Then Stafford led Hayes to a discussion of what DDT does in the human body:

Q: Doctor, physiologically, what happens to DDT when it is taken into the human body?

A: Well, like any other drug or chemical of any kind that is absorbed, it is distributed to all tissues; and the concentration it reaches in different tissues depends on the dosage and also on the chemical nature of the tissue. Because DDT is chemically remarkably inert and because it is highly soluble in fat, it is stored in fat to a much higher degree than in other tissues. This is particularly true following repeated doses, which give an opportunity for this accumulation in the fat to occur.

[DDT] is also and at the same time metabolized or broken down into simpler—and it turns out more water soluble—materials which can be excreted. And so as soon as it's absorbed, some excretion begins. The excretion is—the metabolism is—of, I should say, moderate efficiency. As with other materials of this sort, there is some accumulation in the tissue following repeated dosages.

Now because the rate of metabolism, the rate of excretion [is] related to the concentration in the body and not directly to the dosage, . . . the accumulation in the tissues increases rapidly at first and then more slowly, and gradually reaches a steady state, because the concentration in the tissue has now reached a point high enough to determine a total excretion of as much material per day as is taken in per day.

Q: Are you saying then, in the general population, the amount of DDT stored by the average human being has reached or does reach a plateau?

A: That is true. . . .

Hayes began talking about the amounts of DDT stored in humans and how the amounts had actually decreased in the past few years, after sharply rising for the first few years after the compound was introduced into widespread agricultural use. He then talked about the Michigan coho salmon, impounded by the Food and Drug Administration during the hearing recess because of DDT levels averaging 19 ppm, and said that this concentration would not be harmful. As evidence of this, Hayes stated that the factory workers he had studied had a higher DDT intake; therefore, even if inveterate fish eaters ate 2.2 pounds of salmon per day, they would be in no danger. Hayes continued by examining data on DDT levels in Wisconsin fish, reiterating that, in his opinion, no matter how much was eaten, consumption of fish containing DDT at present levels would not harm human beings.

After this scientific interlude, Stafford tried to bring up the matter of DDT's efficiency in controlling disease, and the possible effect a ban of the substance might have on the underdeveloped countries where great amounts of DDT are still being used. However, Yannacone successfully objected at almost every turn, claiming either that this was a political matter or outside Hayes's competence.

Shortly after the laborious infighting over insect-caused diseases was over, Yannacone began his cross-examination. He quickly set a tone which he kept throughout Hayes's ordeal.

Q: Dr. Hayes, do you really consider yourself the world's foremost toxicologist?
A: That isn't really a question for me to answer.
Q: Oh, Doctor, that is a question for you to answer. I want an answer.

Naturally, Stafford objected, and naturally, Van Susteren sustained the objection.

Arguments, which flared bitterly during the Hayes cross-examination, began almost immediately, starting over the definition of a qualified toxicologist. Hayes claimed he knew of no qualified toxicologist who differed with his views. Of course, this brought bellows of disagreement from Yannacone, and the squabbling continued uninterrupted until the Hearing Examiner interjected: "Both Counsel sit down and be quiet." In the days before Judge Hoffman, a Hearing Examiner's wishes were obeyed.

Yannacone next brought up the subject of safety and dosage to which Hayes replied:

Dr. Hayes: Well, in toxicology it's generally recognized that safety has to be determined in terms of dosage; because there is no material which is not dangerous when given in excessive dosage; nor, in fact, any material, no matter how toxic, which is not safe if given in sufficiently small doses.
Mr. Yannacone: Oh, really?
Dr. Hayes: This is the way that toxicologists find this thing. Now it's in total disagreement with the lay understanding. The lay understanding is that there are toxic materials and then there are safe materials. It turns out that either one can be shifted depending on the dosage. And so the way of finding whether the material is safe is first to test it in animals. . . .

Yannacone didn't like this answer and moved to have it struck from the record as unresponsive, which started another round of legal squabbling. When this squabble was squabbled out, Yannacone took up the validity of Hayes's research and began asking him if he had performed various medical tests on his volunteers and factory workers in an attempt to get Hayes to admit that his studies would only recognize gross neurological or clinical symptoms. However, Yannacone could never trap Hayes into making a flat admission of this, though subsequent cross-examination seemed to point strongly in that direction.

In conjunction with this line of attack, Yannacone next began what was to be his cutest trick at the Madison hearings. Picking up a medical manual, he began reading possible tests for blood abnormalities, asking whether each had been performed. It proved both confusing and embarrassing for Hayes.

Q: Did you measure hemoglobin?
A: Yes, we did.
Q: Did you determine the hematocrit?
A: Yes, sir.
Q: Did you determine the erythrocyte sedimentation rate?
A: I think we did. . . .
Q: Did you determine the amount of C-reactive protein?
A: No, we certainly did not.
Q: Did you determine whether or not there were any sickle cells?
A: Yes, this would have shown up in the differential count. . . .
Q: Did you determine the mean corpuscular volume?
A: I don't think this was specifically measured. . . .
Q: Did you determine the mean corpuscular hemoglobin?
A: Again, I don't think this was expressed. . . .
Q: Did you determine the coagulation time?
A: No, sir, I don't believe so.
Q: Did you determine the clot retraction time?
A: No, but you see, the people had no---
Q: Just answer my questions. You can elaborate a little bit later.
     Did you determine the bleeding time?
A: No, sir.
Q: Did you determine the fragility of red blood cells?
A: No, sir.
Q: Did you do any blood grouping?
A: Why, of course not. It has nothing whatever to do with this subject.
Q: We will get to whether it does or does not in a moment. Did you make Coomb's test?
A: No, sir.
Q: Did you take any blood cultures?
A: No, there was no indication for blood cultures.
Q: All right, just yes or no and [when] we get to the end, sum it all up. Did you determine the blood sugar level?
A: In one study, yes, we measured sugar and found no abnormality. . . .
Q: Did you determine the plasmic prothrombin time?
A: No, sir.

And so the questions continued, to the delight of the pro-petitioner hearing room audience and the chagrin of "No, sir" Hayes and the Task Force lawyers.

After exhausting the manual, Yannacone began questioning Hayes as to exactly what his studies determined about the health of the "victims."

Q: . . . In determining the condition of the volunteers at the beginning of the experiment the only things reported were the ages, the weights, the red blood counts, the hemoglobin level, the white blood cell counts, the polymorphonuclear . . . leukocytes, heart rate at rest, heart rate at exercise and at rest, the systolic blood pressure, the pulse pressure at rest, the systolic blood pressure at exercise, pulse pressure at exercise, the plasma cholinesterase, and the DDT and DDE stored in the fat. Is that a complete statement of the examination tests performed on the volunteers at the beginning?

A: No, there were some other studies which---

Q: Were they reported?

Mr. Stafford: Let him finish his answer.

A: I don't really know. We also tried to follow up as new findings came out. . . .

The probing continued:

Q: All right. Now, Doctor, back in 1956 I think it was when you said you did these studies, did you have a particular experimental reason for conducting them?

A: Yes, we certainly did, and the others too. We were interested in finding a high [DDT] level that was safe. And we estimated that the one we tested would be safe. We were also interested in studying [DDT] storage in relation to dosage and, if possible, excretion. Certain improvements in the method were developed during the first study and carried out in much greater detail in the second, with literally thousands of chemical analyses on urine in the second study.

Q: Those were your experimental reasons?

A: Those were the reasons.

Q: To study storage and excretion of DDT and its metabolites in humans, right?

A: That's correct.

Q: And those experimental objectives were reached by tests of the various excretory products of the human subject, weren't they?

A: As well as we could with those methods.

Q: And you made some determinations of the level of storage and excretion, right?

A: Yes, sir. . . .

Q: All the other medical testing was peripheral, wasn't it?

A: Why, no, we were looking for any possible effect, with emphasis on those that had been seen both in man and in animals as a result of exposure to this compound.

Q: Well, what was the purpose of the studies now, to determine and measure the excretion and storage of DDT and its metabolites?

A: And to see if we could in fact find any effect from the particular dosages that we had chosen to use.

Q: Any effect, or only certain effects?

A: Well, we were looking for any effect. But the tests emphasized those that had some chance of being positive as indicated by what was known.

Q: At that time?

A: At that time.

Q: In other words then, Doctor, your tests were somewhat predetermined by your expectation of what might occur from what you knew at that time?

A: I believe this is characteristic of all tests; they're designed in terms of what is known.

Yannacone's response to this was merely and cynically, "Oh?"

Then after a series of questions on the circumstances of the experimentation done on prisoners by Hayes in the 1950's, Yannacone threw out a very significant query.

Q: Now, Doctor, at that time was any attempt made to measure the enzymatic functions of the liver?

A: . . . In answer to that question, I think your answer is no. We actually planned before I retired from the service—and the work will go forward—studies of drug metabolism in people with heavy exposure to DDT. . . . It has, in fact, been done in Sweden, and this reveals an effect which not only we but I believe no one was prepared to look for at the time this first study was made.

In the light of the testimony of witnesses for the petitioners on the hepatic enzymatic-inducing activities of DDT and its metabolites, this was an important admission of omission.

Yannacone then began questioning the statistical methods of analysis used by Hayes in his work, first impugning his knowledge of statistical methodology; then stating that Hayes, because of the crude nature of his analyses, might have omitted significant items from his published papers; and finally zeroing in on his use of the t-test* as a primary statistical tool for the analysis of data, stating that its measurements weren't sufficiently sensitive. To illustrate this, Yannacone began a line of attack, memorable to all and infuriating to the pro-DDT forces.

Q: Doctor, in the course of your medical studies did you ever have occasion to hear about the substance commonly referred to as testosterone?

---

*a statistical test used to compare the mathematical means of normal populations which have unknown standard deviations

A: Yes.

Q: Are you married, Doctor?

A: Yes, sir.

Q: Got any children, Doctor?

A: Yes, sir.

Q: Have you got any idea what the current levels, blood levels, of testosterone in your system were about the time you were fathering those children?

A: No, sir, I haven't the faintest notion; but they were adequate.

Q: Do you know how little difference there is, Doctor, between "adequate" and "inadequate"?

A: No, sir, I don't. . . .

Q: Doctor, I want you to assume that the working level of testosterone sufficient to maintain your male functioning is something on the order of five parts per billion, and that [a] level below two parts per billion is sufficient to get rid of your mustache, the hair on your chest, the hair on a number of other places, raise your voice a few notches, and probably preclude your fathering anything other than freak shows. Now, Doctor, assuming that difference, I want you to sit there and—if you want to use pencil and paper you may—make the calculation of the level of difference between two parts per billion—that's two times ten to the minus ninth—and five parts per billion—which is five times ten to the minus ninth. Now, if my arithmetic is correct—and you may check me with pencil and paper—that difference is three times ten to the minus ninth, or three parts in a billion parts.

Now, Doctor, can you now tell us that the student's t-test is sufficient to measure those kinds of differences in living populations? Yes or no?

Another legal battle erupted over this question, but the question itself was enough to inflict damage on Hayes and didn't need an answer.

Shortly after this exchange, the witness was excused from the stand in order to accommodate Harry Hays. Wayland Hayes was recalled two days later, and Yannacone began questioning him again on the subclinical effects of DDT, especially those related to enzyme induction.

In answering Yannacone's questions, Hayes returned to old ground: the search for overt neurological signs as symptoms of DDT poisoning. And agile Yannacone just as quickly returned to his argument that the neurological studies that Hayes and his colleagues had conducted were insufficient.

Mr. Yannacone: Were any procedures utilized to determine the subclinical effects of enzyme induction?

Dr. Hayes: No, this was not done. What we did, for example, . . . in the neurological examination we went through [was] this. One of the most sensitive tests for the effects of DDT is tremor. It is entirely possible that a person might have tremor, from whatever cause, without realizing it, without thinking of himself as ill and without being inconvenienced. We examined these men, for example, for tremor. . . .

Q: What tests did you employ to determine whether there was tremor?

A: Just the usual test of having . . . the man extend his hands, and to look at them and see if there was tremor. . . .

Q: You were not looking for tremors which would require determination by more sophisticated tests, were you, Doctor?

Yannacone then went back to the subject of the liver and microsomal enzyme induction. Hayes agreed that such induction occurred, but denied that any data was available which showed that the induction of enzymes by DDT was harmful, backing up this assertion with a detailed description of liver function. After this, Yannacone changed course, returned to the nervous system, and began questioning Hayes about the effects of DDT on nerves themselves.

Q: Doctor, do you know, or can you tell us what you believe is the mechanism of action of DDT on the nervous system?

A: Neither I nor anyone else know the mechanism of action of DDT on the nervous system on a biochemical level. . . .

This answer was to be rebutted, at least in part, later on in the hearing by Dr. Alan Steinbach, a neurophysiologist, then teaching at Albert Einstein College of Medicine who was to be one of the most persuasive of all the petitioners' witnesses.

But Yannacone was far from finished with Hayes. In the midst of what often appeared to be merely repetitious questions, Yannacone would throw in a bombshell. For example, during the answer to one of Yannacone's questions, Hayes stated that damage to a single cell was not significant to the human body. Yannacone pounced.

Q: Doctor, is it your opinion today, sitting here as an M.D. and a professor, in 1969, that it is possible to damage even a single gene of a human chromosome in a human egg or sperm cell and not run the risk of producing serious damage, ultimately, to the organism that might survive birth?

Hayes's answer was that usually damage to a sperm cell or egg cell produced sterility, but Yannacone had made his point and made it well. He continued.

Q: Doctor, I'm not interested in sterility. I'm interested in *mutagenesis*. Have any studies to your knowledge been done on whether

or not DDT has any mutagenic effects whatsoever on any organism?

A: . . . These studies have been made, and they have been negative,
to my knowledge.

However, later on that year, information was released which
strongly indicated that DDT was not only mutagenic but *carcinogenic*
as well, and this information ostensibly led the federal government
to make the long-delayed ban on DDT.

Yannacone then moved back to the area of human diseases such
as rheumatoid arthritis and heart disease, not easily detectable in their
early stages, attempting to show that DDT, too, was possibly having
effects that were not yet seen; and from there to the work of Dr.
William Deichmann. This Florida toxicologist had performed a series
of autopsies on people who had died of terminal diseases and had
discovered that their bodies contained significantly higher concentrations of DDT and its metabolites than normally found in healthy
individuals.

Q: Doctor, Dr. Deichmann reported,* did he not, that in individuals
who died of various terminal diseases, total DDT in body fat was
found to be elevated twice, two and a half times in arteriosclerosis,
leukemia, carcinoma, hypertension, and encephalomacia? . . .

A: Yes, he did. And as I have already pointed out, his conclusion
was that he could draw no conclusion about the relation of
causation from data at hand. . . .

Yannacone then caustically asked:

Q: Doctor, you are telling us, are you not, that Dr. Deichmann's
proper scientific statement that he cannot positively relate these
findings to the level of DDT and say that DDT is the absolute
direct single cause thereof, is equivalent to your statement that
you can now sit in 1969 in that witness chair and tell us in the
face of Dr. Deichmann's paper that you are a hundred per cent
medically sure that DDT is absolutely safe? Doctor, is that what
you are telling this hearing?

A: Yes, sir, this doesn't interfere whatever with reaching that conclusion. . . .

Q: And you still feel in spite of [Deichmann's data]—and it's 1969
data—you are still sure that DDT is a hundred per cent safe, is
that what you are telling us?

A: That's correct.

---

*J. L. Radomski, W. B. Diechmann, and E. E. Clizer, "Pesticide concentrations in the
liver, brain and adipose tissue of terminal hospital patients," *Food and Cosmetics
Toxicology* 6 (1968): 209–20.

Yannacone had one more important thing to bring out with Hayes, suggested to him by Dr. Goran Lofroth of Sweden on his arrival in Madison. This was a matter discussed at the Swedish conference which had taken place during the Madison hearing recess and had led to the banning of DDT in that country.

Q: Now with respect to the Laws study,* there weren't any women involved in that study, were there?
A: No, sir.
Q: There weren't any infants involved in that study, were there?
A: No, sir.

Yannacone asked Hayes the same question about the other studies on which he had based his testimony.

Q: Now, Doctor, there were no infants or children in that occupational study or in your convict study, were there?
A: No, sir.

After this, it got emotionally hectic again in the hearing room but Yannacone had made his point: the efficiency of liver detoxification mechanisms might vary with age and sex, yet all of Hayes's opinions were predicated on data from adult males.

But Hayes still stuck with his opinion that DDT was safe.

Yannacone's brutal cross-examination of Wayland Hayes was but a single technique among the tactics used to discredit the DDT industry's toxicologist and to establish the potential danger of DDT to humans. As the parade of witnesses continued—Richard M. Welch, biochemical pharmacologist from Burroughs Welcome and Company; Theodore L. Goodfriend of the University of Wisconsin School of Medicine; Goran Lofroth of the Royal University of Stockholm; and Alan Steinbach of the Albert Einstein School of Medicine—other tactics appeared.

Welch, who had testified before the recess, had been given the job of carrying the story of the enzyme-inducing qualities and hormone-like effects of DDT to a species more physiologically akin to man, the rat. By introducing this research on higher mammals, Yannacone hoped at least to lessen the impact of Hayes's scene-stealing human work.

Dr. Welch: . . . We initially studied the effects of chlordane . . . and DDT on the metabolism of testosterone in . . . rats. Now testosterone is a naturally occurring *androgen* that's present in man

*E. R. Laws, Jr., A. Curley, and F. J. Biros, "Men with intensive occupational exposure to DDT," *Archives of Environmental Health* 15 (1967): 766–75. This study was conducted on men working in DDT manufacture at the Montrose Chemical Company and was the basis of much of Dr. Hayes's testimony.

and in rodents. We found that upon the daily administration of DDT or chlordane there was a marked increase in the rate or the metabolism of several steroids* by liver enzymes in the rat. . . .

Several drugs and insecticides belong to a group of compounds now generally recognized to be enzyme-inducing agents, that is, they will cause an elevation in the enzyme level in the particle of the liver called the microsome. These enzymes are collectively called a mixed function oxidase system with a broad spectrum of activity, that is, they are rather nonspecific in that they metabolize, in addition to steroids, very many drugs and foreign compounds. When one administers a compound that causes an elevation in these enzymes, it is referred to as an enzyme-inducing compound.

Well, further studies with DDT and chlordane revealed that, with respect to testosterone metabolism, we got marked changes in the degree of hydroxylation of testosterone. . . . [Also,] the ability of DDT administration in vivo to stimulate the metabolism of estradiol in vitro by these liver enzymes suggested that these insecticides might decrease the biological activity of estrogen in animals. Now in vivo ---

Examiner Van Susteren: Before you go any further, while probably most of those in this room understand what you mean by "in vivo" and "in vitro," would you explain for the record what is meant by both terms.

Dr. Welch: "In vitro" means the addition of something into a test tube, [the addition] takes place outside of the body; while "in vivo" is occurring within the animal.

Now, as I just said, indicating that the ability of DDT to alter the metabolism of estradiol and other steroids in vitro suggested that perhaps there was an alteration of enzyme activity and a decrease in biological activity in vivo.

We proceeded to investigate several insecticides. And we have found that indeed something like chlordane will cause a marked decrease in the action of estradiol in vivo. . . .

When we proceeded to do the same type of thing with DDT, because DDT was active in vitro, we found some rather interesting results. We found we couldn't do the experiment in the animal because DDT had some action on the uterus itself. . . .

We noticed upon the injection of DDT we got a marked increase in the uterine wet weight at six hours after the administration of DDT. This effect suggested to us that perhaps DDT itself did possess some estrogenic activity, that is, it acted in some capacity like the natural hormone estradiol would act.

Examiner Van Susteren: Well now the Examiner is getting curious as to what type of a test subject you did your work on.

---

*among them, testosterone

Dr. Welch: We studied immature female rats, for good reason. Immature female rats would not be able to make any estradiol of their own and therefore would serve as a good test system for evaluating estrogenic compounds. . . . In addition, we have demonstrated this [increase in uterine wet weight] to take place in adult ovariectomized rats where the level of estradiol is low. . . .

Examiner Van Susteren: And by "ovariectomized" you mean where the ovary was removed?

A: That's right. The ovary would complicate this enzyme system, because it would make estradiol and influence the uterus. And since we were trying to evaluate the effects of DDT alone on the uterus, it was necessary to remove that organ which was responsible for the endogenous synthesis, that is, in vivo, of estradiol. . . .

Now, an increase in the uterine wet weight is not the sole criteria for determining whether a substance does possess estrogenic activity. There are many biochemical events that take place in the uterus following the administration of an estrogen, and we proceeded to explore some of these.

What generally happens is that there is an increase in the incorporation of *glucose* into uterine lipid-protein *ribonucleic acid* and in the acid-soluble constituents of a rat uterus following an [injection of] estrogen. Since this is known to take place, we investigated the effects of DDT on these biochemical parameters. And, indeed, we did find that DDT did cause an increase in the incorporation of uniformly labeled glucose into these various fractions of the rat uterus.

Further studies from our laboratory indicated that DDT, . . . having estrogenic properties, would indeed compete with estradiol for receptor sites on the uterus, that is, [DDT compounds] would prevent the combination of estradiol with a receptor site on the uterus virtually because they themselves would occupy that receptor site. . . .

Mr. Yannacone: Now, Dr. Welch, in the course of your research, what is the smallest amount of DDT that you found produces estrogenic activity?

A: . . . The administration of a single dose of five milligrams per kilogram of DDT (tech) caused a statistically significant increase in the uterine wet weight; while the administration of one milligram per kilogram of the o, para-prime DDT analog caused a significant increase in the uterine wet weight.

Those are the lowest levels that cause the effect. . . .

. . . We have not done a dose response curve on how little DDT is necessary to cause changes in the liver enzymes of rats. However, there is adequate literature on this aspect of it. . . .

After the usual argument over whether scientific literature could be introduced into the record, Dr. Welch continued.

Dr. Welch: In determining . . . what level of DDT can cause biochemical changes in the liver of rats, I indicated . . . that in a
publication submitted by Hart and Fouts in *Toxicology and Applied Pharmacology*, Volume 5 in 1963, they found that five parts
per million of DDT when given to rats for three months caused
an increase in the activity of enzymes in the rat liver.

Welch then cited another source which stated that the DDT level
could be reduced to 1 part per million of DDT, and a final source
stating a level as low as 40 micrograms administered for four weeks
would be effective.

Mr. Yannacone: Now Doctor, would you summarize for us briefly what
conclusions you have drawn with a reasonable degree of scientific
certainty in your professional capacity as a biochemical pharmacologist with reference to your work and the other work that you
have described here, in particular relating the effects of DDT on
biochemical and biological systems?
A: Well, the exact relevance requires more study. . . . But in a publication by Swabe* in Germany they have indicated that a fat
residue of DDT of 10 parts per million in rats causes a change
in the pharmacologic action of pentobarbital, a commonly used
drug in man; and they have indicated that this change in pharmacologic activity is correlated with an increase in the ability of the
liver of the rats to metabolize pentobarbital. . . .
Examiner Van Susteren: It inhibits the pharmacologic activity?
A: Right, because it shortens the duration of action of pentobarbital.
The explanation for this is that this amount of DDT increased the
enzyme level in the liver of the rats responsible for the breakdown
of this drug.
   Now this 10 parts per million is within the range of DDT found
in human fat. . . . Thus, if one can extrapolate data from animals
to man then one would say that a change in these enzymes
probably does occur in man. . . .

Bringing rats even closer to humans, at the end of Welch's direct
testimony Yannacone asked:

Mr. Yannacone: Now Doctor, the steroid hormones testosterone,
estradiol, [and] estrone that you described in your work, these
were found in rats, were they not? . . .
A: Testosterone and estradiol and hydrocortisone are known to be
present in the rat.

---

*Swabe, "DDT—Speicherung bei der Haltung Von Veruchstiern alsmogliche Fehler-
guelle bei Arzneimittel—prufungen" *Arzneimittel-Forsch* 19 (1964): 1265.

Q: All right, now the substances that you described as being present, these substances that you found in the rats, they are no different than the same named substances found in other mammals like humans, are they?

A: That's right.

Q: And for this reason, rats are commonly chosen as laboratory animals, are they not?

A: That's right.

Q: All right. In other words, the testosterone you found in a rat is the same as the testosterone you find in a human?

A: Yes.

Q: The estradiol you find in a rat is exactly the same as the estradiol you find in a human, is that right?

A: That's right.

Mr. Yannacone: Okay Doctor, thank you very, very much. I have no further questions.

With S. Goran Lofroth, Yannacone's strategy was much more devious. Lofroth was a top European environmental scientist: he had been Chairman of the Working Group of Environmental Toxicology, a committee appointed and funded by the Swedish Natural Science Research Council to make an exhaustive search of the literature on man's burden of DDT compounds and the effects of those compounds on other mammals. In addition, Lofroth had participated in the Swedish conference which had led directly to the banning of DDT in that country. A crucial fact here was that Wayland Hayes, too, had been at that hearing and by all accounts had "lost," a fact which had not made the American press and one that Yannacone was most anxious to introduce. With this background, Lofroth should have been one of Yannacone's strongest witnesses; the fact that he wasn't and the reasons for that fact, form one of the more bizarre stories of the Madison hearing.

Lofroth, rather than appearing as Yannacone's witness, took the stand as the "impartial" witness of Robert McConnell, the Public Intervenor. Now, McConnell's position was a curious one, very much analogous to that of a Swedish ombudsman. He represented the public's interest in matters directly concerning them where they would otherwise have no representation. Officially, McConnell was objective in the matter of DDT's innocence or guilt.

After establishing Lofroth's academic qualifications, McConnell in one motion immediately attempted to introduce as exhibits the 56 scientific papers that Lofroth had brought with him from Sweden to serve as a basis for his scientific opinion. This attempt failed because of repeated objections by Stafford to their admissibility. Nonetheless, led by McConnell, Lofroth grimly continued. He began by giving a summary of the average concentrations of DDT found in man, concluding with the statement "the average concentration in the human adipose tissue in the whole world—[and] this is just the average of the whole world population—seems to be in the range of 10 to 15 ppm." Lofroth then carried his survey to the concentrations of DDT compounds found in human milk. As historical background, he cited an early report by Woodard published in *Science* in 1945 stating that in experiments with dogs, ingested DDT is indeed excreted in milk. From that early finding Lofroth continued on to reports by Lang et al. on the presence of DDT in human milk; to a report in the *British Medical Journal,* 1965, that the sum of the DDT compounds in the human milk of 19 sample subjects averaged .128 ppm; to a report by Quinby et al. published in *Nature* in 1965 stating an average figure of .17 ppm for subjects in the U.S.; to further findings in the Soviet Union.

At this point the cold war, supposed in most quarters to be an issue of the 1950's, popped in with Stafford's objection:

Mr. Stafford: Your Honor, reluctantly I'm going to expand my original objection and ask that the witness be instructed to make no reference whatsoever to DDT and its metabolite residues in the Soviet Union, China, or any of the iron curtain countries without first presenting in this record a full foundation as to the accuracy of these residues, how they were taken, and how the DDT presumably was manufactured in these countries. . . .

The objection was overruled, to the laughter of most of those in the hearing room who were hoping for better East-West relations; and Lofroth proceeded, still without the benefit of his supporting documents, to give the opinion that was a high-point of the petitioners' case.

Dr. Lofroth: One arrives at the conclusion that the average concentration of DDT compounds in human milk is about .1 to .2 ppm. The average daily intake of human milk of a breast fed baby is about 150 grams milk per kilogram body weight. Easy calculation means that the daily intake of DDT compounds is about . . . .02 milligrams DDT compounds for the average baby.

The World Health Organization and Food and Agriculture Organization of the United Nations make, jointly, recommendations about acceptable daily intakes of pesticides for man. For DDT compounds, that is DDT plus DDE plus DDD, this is .01 milligram per day [per] kilogram body weight.

Incidentally, this DDT plus DDE plus DDD are what are generally called DDT compounds.

Thus breast fed children ingest about twice as much DDT compounds as the recommended daily maximum intake. Of

Before it is shown that DDT compounds are safe for man, shown with a significant degree of scientific certainty, one should not further spread DDT in the environment.
S. Goran Lofroth

course, this is the average and some ingest less and some ingest more.

My opinion is that from these last two statements it is obvious that many breast fed children ingest more than this recommended maximum daily intake.

Furthermore, it is in the range of exposure to which laboratory animals show pharmacodynamical changes. What these changes mean is not known, [so] one cannot predict the consequences with similar or other changes at work in man, and one does not know what the future might bear for mankind.

Thus, before it is shown that DDT compounds are safe for man, shown with a significant scientific certainty, one should not further spread DDT in the environment. . . .

Mr. Stafford: While there's a moment here, I wish to move the witness' testimony regarding safety to mankind or to babies . . . be stricken. . . . There's been absolutely no foundation in his background to qualify him to state an opinion in this regard.

Examiner Van Susteren: Just a moment. He merely used the information that he cited and that he read, and the studies, incidents thereto in the formulation of his opinion.

Mr. Stafford: He has no independent opinion, then, I presume?

Examiner Van Susteren: He has rendered his opinion.

The admissibility of Lofroth's documents was not the only problem to plague his testimony. Soon arguments arose as to whose witness Lofroth actually was, Stafford claiming that, since Lofroth's transportation expenses had been paid by the Environmental Defense Fund, he was, in fact, Yannacone's witness; and Van Susteren seeming genuinely confused by the whole situation. The court record, reminiscent of Pope's famous couplet:

> I am his highness' dog at Kew
> Pray tell me, sir, whose dog are you?

contained ample evidence of this time-consuming battle in its 145 pages, but recorded particularly bitter arguments when Yannacone tried to "cross-examine" the witness. Naturally, Stafford insisted that Yannacone was not cross-examining at all, but instead, was merely continuing McConnell's attempts to get Lofroth's disputed documents on the record—and, indeed, Yannacone did introduce the 13 most important ones.

Mr. Stafford: Now what's happened is that Mr. McConnell called his witness, had some difficulty qualifying him, as I think is clear in this record, and asked him some opinions, and he rendered some. Clearly Counsel for the petitioner has a right to cross-examine this witness, and that examination must be restricted to the witness' direct examination. I'm objecting because, of course, what's happening here is that the petitioner is attempting to shore up the rather gaping holes of these denied judgments. . . . Now this is not cross-examination; it's a transparent device to give Counsel two kicks at the cat now, and this is improper.

The various Lofroth battles culminated during Stafford's final examination of the witness, an examination which, at last, gave Lofroth the opportunity to make a significant point, but also gave Yannacone the opportunity to enact one of his most amusing scenes.

Mr. Stafford: When was DDT first generally used in the world, if you know?
Dr. Lofroth: 1942. . . .
Q: And there've been approximately 27 years go by now. Do you agree that as far as human children are concerned, sufficient time now has gone by so that at least two generations of human children could be affected by the DDT in diet or atmosphere?
A: No. . . . The large scale use started a little later than that so my opinion is that the majority of the population has yet only been exposed to DDT, say, something between 20 and 25 years. . . . No one above this age was exposed to DDT in the very young age at the time when they are sensitive to poisons, that is, at the

stage when the liver hasn't developed the [de]toxification mechanism. . . .

Q: Yes, well, there are many, many breast fed babies who were exposed
to DDT at the breast of their mothers, many years ago who are
now adults, are there not, in the human population? . . .

A: There must be.

Q: Why, of course. Now do you know of any instance or episode
where these children . . . have been harmfully affected by DDT
from their diet or from the environment? . . .

A: *To my knowledge there has been no investigation on the thing
even. That's even worse.**

Shortly thereafter Yannacone raised an objection to a line of questions used by Stafford. Examiner Van Susteren responded to his objection by saying:

Examiner Van Susteren: All right, now just a moment. The Examiner
wants to point something out here. He stated yesterday and this
morning that he had serious and grave doubts as to whose witness
Dr. Lofroth was. If you had cross-examination this morning, are
you trying to make him your witness now? . . . It's becoming
increasingly obvious today that Dr. Lofroth appears to be not only
Mr. McConnell's witness but yours.

Mr. Yannacone: Mr. Examiner, I object to that.

Examiner Van Susteren: Any differentiation between cross-
examination and direct examination here turns into a farce. Now
gentlemen---

Mr. Yannacone: Mr. Examiner---

Examiner Van Susteren: Let's get on with this.

Mr. Yannacone: For the record, let the record show I leave the room
now for the balance of the examination of Dr. Lofroth. He is Mr.
McConnell's witness and I am sure that if the Examiner will let
him answer the questions, he can answer any question Mr.
Stafford might like.

Mr. Stafford: Good bye.

—Or almost "good bye." Yannacone simply couldn't resist popping
in once more, occasioning a closing quip from Examiner Van Susteren:

Examiner Van Susteren: The record may show that, while Mr.
Yannacone was absent during the examination of Mr. Stafford and
Mr. McCallum [another Industry lawyer] he has his trusty tape
recorder in front of him and that the tape recorder was watched
by the Examiner and never once stopped.

---

*italics, editorial

Throughout the DDT litigation, Yannacone had been pressing for a neurophysiologist to testify about DDT and its effect on the nervous system. Wurster and others, however, insisted that no data from neurophysiology with all its sophisticated mathematics was needed; that the whole case could be based on the story of thin egg shells and calcium metabolism in birds. But Yannacone kept saying, although "bird nerves and people nerves are essentially the same, people don't lay eggs with shells."

So Alan Burr Steinbach, a 28-year-old neurophysiologist at the Albert Einstein Medical School was found, one of a "new breed" of scientist increasingly evident, equally at home discussing science or the situation in the urban ghettos. Here was a witness who could break the backbone of the DDT Task Force contention that the petitioners could present no data that was really relevant to man for, by the time Steinbach was through testifying, everyone in the hearing room had a basic understanding, not only of the effects of DDT on nerves, but of a possible physiological basis for the tremors George Wallace had observed in birds and of the history of neurology as well.

In many ways, Steinbach was the most impressive witness that the petitioners corralled. He managed to take the subject of neurophysiology, one alien to most people in the hearing room, and make it relevant to their lives and to the contaminants in their environment.

Mr. Yannacone: Now, Doctor, in the regular course of your professional activities, have you had occasion to investigate the effects of DDT on the nervous system?

Dr. Steinbach: Yes.

Q: Now, Doctor, before we come to the opinions, conclusions, and whatnot, would you briefly outline for the record the mechanism of nerve conduction?

A: I think that in order to explain it fully, it would be well to give a small historical background, if it would be all right.

Q: Go ahead.

A: Basically the place to start is way back at the beginning with Galvani, [in] 1786, who discovered what he called animal electricity; that is, that a frog or other animal, if you touch the nerve to a piece of metal or even touch the nerve to another frog, you could get a contracton in the muscle fiber. Galvani developed a theory based on the idea that the nerve was a tube with a surrounding insulating layer and a sort of fluid internal layer, and that the fluid internal layer in some way carried what he called at that time "electrons," although that is not the same as the present term electrons, . . . and that somehow as a result of the flow of electrons down the nerve, one induced a contraction in the muscle fiber.

This was immediately taken up and seized on by Volta, who was at that point rather high up in the Academy of Sciences, and it was probably the first case of a well-meaning but perhaps academically not too qualified scientist being squashed by a superior position, but not superior reasoning. In any case, Volta successfully discredited Galvani and it wasn't until about, say, 1840, that Galvani's ideas were proved to be substantially correct.

Steinbach then took his audience through the nineteenth-century world of science, tying in Matteuchi's experiment showing that a healthy nerve has more electrical potential than an injured one; Michael Faraday's mathematical calculations that led to the invention of the galvanometer, the instrument which measures the extent of electrical current flow in the nerve; and Golgi's efforts to prove the existence of junctions between cells.

After tracing the history of neurology into the twentieth century, including the experiments which led to the determination of how long it takes for a nerve impulse to pass along a nerve, how strong that impulse is, and how the electrical impulse is generated, Steinbach laid the groundwork for introducing the Hodgkin-Huxley equation, the mathematical formulation that was to be the basis of his substantive testimony. Yannacone, who is enamored of technology and its terminology and electronic gadgetry had Steinbach go into minute detail in describing the methods and modes of deriving this equation, introducing into the record fourteen pages of complex mathematical testimony, incomprehensible in all but general outline to most of those present. Finally, in summary, Steinbach said:

Dr. Steinbach: In order to check the accuracy of our description of the ionic process [which occurs within a nerve we] should be able to use this [Hodgkin-Huxley] equation with the constants, and mathematically generate a time-variant function, that is, an action potential which will match exactly the action potentials that one actually records in the nerve. And this was the successful test of the Hodgkin-Huxley equation, . . . that they were in effect able to successfully match the performance of the nerve with a solution of an equation generated from their description of the process. . . . What one has in sum-up of this is an empirical description of the way in which the nerve functions. One has perhaps the most sophisticated tool for the investigation of neurophysiological and neurological processes that could have been hoped for considering the instruments available at this time. . . .

With this analytical tool, since 1952, neurophysiologists have proceeded to attempt to investigate the mechanism of action of various chemicals that have been thought in the past to interfere with nervous system function.

So, at last, DDT entered the picture.

Mr. Yannacone: Doctor, in the course of your regular professional activities have you had occasion to investigate the effects of DDT upon the nervous system?

A: Yes, I have.

Yannacone then threw in a few qualifying questions before getting down to the business at hand.

Q: Now, Doctor, can you—just yes or no—can you form an opinion with a reasonable degree of scientific certainty as to whether or not DDT has any effect on the nervous system?

A: Yes.

Q: Doctor, does DDT have any effect on the nervous system?

A: Yes.

Q: Doctor, can you describe for us the mechanism of action of DDT upon the mammalian nervous system with respect to [its] conductance mechanisms? . . .

A: DDT was examined on various invertebrate nervous systems starting in 1946. . . . The experiments basically were done, first with the cockroach and then with crustaceans. The observations were quite definite that DDT did have an effect.

In the case of the cockroach (the insect), and the crustacean, the effects occurred at very low concentrations. The effects were variable. In most cases in the cockroach, it consisted of repetitive firing in the nerves, where one impulse applied to the nerve no longer evoked a single message going down the nerve, but rather a large volley of messages. The behavior of the cockroach, at the same time, showed disorientation, running about, tremoring, kicking of legs in the air, and death. The idea was the DDT, causing repetitive firing, jammed all the transmission lines, overworked the cockroach, and caused death due to exhaustion.

In the case of the crustacean, repetitive firing wasn't so much an aspect. Instead, one saw a change in the afterpotential following the nerve impulse. Based on the formulation in the Hodgkin-Huxley equation, the afterpotential is due in part to the turning on of the potassium conductance and in part to the slowness of the turning off of the sodium conductance. . . .

And then a little later:

Dr. Steinbach: DDT [is] in a class by itself as far as nerve active agents are concerned, in that I have never heard anyone state with confidence they felt there was a lower limit to the effect or a threshold concentration.

Steinbach then stated what was to be the most significant part of his testimony:

Dr. Steinbach: DDT, once applied, doesn't come off within the time course of the experiment. And by that I mean a matter of hours . . . because that was as long as we kept [these experiments going]. This makes it different from any other of the so-called low molecular weight toxins that I know of. Snake venoms, of course, chewing up the membrane, do effectively irreversibly damage it. But the other toxins, curare, veratrine from belladonna, tetrodotoxin from the puffer fish, saxitoxin from the clams, scorpion venoms, black widow spider toxins, are all reversible to some extent. The local anesthetics are reversible. DDT, as far as our experiments were concerned, was not reversible. . . .

A prolongation of the active state of the nerve means—coupled with no change in the inactivation process for sodium—that after one nerve impulse, when the channels become ready to conduct or to open again to conduct a second impulse and examine or test the potential across the membrane, they will find that, according to the potential, they should already go again. That is, no further impulse would be necessary to make [the nerve] fire another signal. This could produce repetitive firing. On the other hand, in other nerves where sodium inactivation takes a longer time, one could simply find that the nerve, on reactivating its sodium mechanism, finds that the potential is already too high to fire . . . and instead [the nerve] would merely remain quiescent. This would produce complete failure of the transmission line or at least intermittent failure. Either of these mechanisms quite clearly and very conclusively could cause tremors and could cause grave disturbances in terms of the ability of the animal to move or to make motions.

I think that the connection between these two is not all a matter of a far-flung difficult connection to make, but a question of saying: Can the . . . data account for the observed biological effects? Yes. Does it prove that they are caused by that? No. But that it can account for it is unequivocally true.

I really think that the observations outlined by Dr. Wallace both today and in his earlier papers, when no one else was interested in this (the effects in animal populations), could definitely be accounted for simply on the basis of the known mechanism of action of DDT on the nerve.

Shortly thereafter, Yannacone asked Steinbach the big question:

Q: Now, Doctor, will you proceed with your opinion as to whether or not DDT can exert sublethal or subclinical effects on the nervous sytem?

A: . . . One does not voltage clamp human beings' nerve fibers for the simple reason [that] they are too small. There's another reason for that, too. Most humans don't like to have their nerve fibers

taken out. But that in every respect in terms of conduction veloc-
ity, in terms of the dependence on temperature, in terms of the
toxicology, and every other respect, differences in terms of chemicals that act such as the ones I have described should not be
great between, say, higher mammals [such as man] and amphibians. One might expect more differences between crustaceans and
amphibians, but it turns out there don't seem to be as many as
one would have thought from actual experimental data in this
case. . . .

In terms of the sublethal doses, I think that there are several
important points about DDT's action. One is that [DDT] . . .
apparently acts—well, for our experiments, irreversibly. I can't say,
you know, whether it ever is reversible. Conceivably it might [be].
But we had no indication it [is]. It certainly acts for a long time.
Presumably this is because of the high lipid solubility.

Second is that in terms of the actual effect, the change in the
turning off phase of the sodium, that this is the type of a change
that very easily would reflect itself to a subnoticeable, not only
sublethal effect. That is, one could easily have a buildup in the
duration of the action potential short of absolute catastrophe
where the nerve would no longer be able to fire action potentials
reliably. The problem with this, if it were a one-shot affair where
a toxin caused this effect for a short time and then went away,
one wouldn't be in trouble. The problem is that DDT seems to
have a long term of action, so that this sort of sub—again, not
subnoticeable—effect . . . is unique to DDT, I think, among the
chemicals I have looked at.

That was the nub of the direct examination of Steinbach. During
the cross-examination Stafford emphasized that Steinbach's work was
not done on live animals, with the young neurologist pointing out
that you can't do this type of experimentation with such subjects.
Faced with self-assurance and logic such as this, Stafford gave up the
game quickly, leaving Steinbach's testimony as an impressive rebuttal
to the gross neurological examinations of Wayland Hayes.

However, there was still a hole in the petitioner's case: with 2664
pages of testimony already on the record, they had yet to find an
honest-to-goodness M.D. to refute the good doctor, Wayland Hayes.
To fill this hole, the petitioners brought in Theodore Goodfriend, an
assistant professor of internal medicine and pharmacology at the
University of Wisconsin School of Medicine.

However, the best Yannacone could get from the doctor in the face
of Stafford's ubiquitous shouts of "Objection" and "Irrelevant" was
an answer to the query:

Q: Doctor, will you give us your opinion as to whether or not it can
be determined from the existing state of evidence whether DDT

is absolutely safe to humans at present levels of body burden and exposure?

A: It's my opinion that based on what I have read and my knowledge of what would be acceptable criteria, that one cannot say that DDT is absolutely safe.

Perhaps a tentative note to end on, as had been many sounding throughout the testimony of the four that Yannacone had called to refute Wayland Hayes. Yet, the public could still say, "Yes, there is doubt as to the safety of DDT" and, in the face of, until then, seemingly insurmountable odds, that was enough.

9

Getting Them Back
on the Farm

A pest is an artificial, nonbiological, totally subjective human label
for an organism that happens to do something we don't approve of.
Dr. Donald Chant

One of the not so subtle charges frequently laid at the doorstoop of
the petitioners at the Madison hearings was that they were, in the
words of one unofficial industry spokesman, "a bunch of goddam
birdwatchers." There was an element of truth in this allegation. For
the most part, those testifying against DDT at Madison were theoretical scientists unconcerned with the practicalities and problems of
agriculture. This was simply brought out by McLean in his cross-
examination of Wurster when he asked the biologist if he had ever
worked on a farm or had any experience with actually controlling
insect pests. His answer was, of course, no.

As fine as the experts testifying for the petitioners were as scientists,
they were all urbanites trying to tell the farmers what to do; instead
of the traditional American legislative problem of rural domination,
the farmers were being threatened by the big city fellers. But Yannacone was shrewd enough to include at least a token force of men
intimately concerned with the day-to-day problems of feeding a
potentially hungry nation. For this reason, perhaps the testimony of
Drs. Robert van den Bosch, Paul De Bach, and Donald Chant, showing
the world that there was an alternative to the widespread use of
chemical pesticides, was among the most significant elicited by the
petitioners at the hearings.

Here were three scientists who plainly and authoritatively explained
to the public, and perhaps even to some farmers, that the bill of goods
being sold to them by the pesticide salesmen was not necessarily the
correct one. In spite of all the high-powered advertising carried by
agricultural trade journals attempting to prove that Brand X pesticide
is all that stands between the farmer and financial ruin, there are ways
of controlling pests whose populations exceed the threshold of economic damage—without poisoning the world.

The most eloquent of Yannacone's triumvirate was Dr. van den Bosch, an ex-advocate of DDT. (His abdication may have seemed to the industry as incongruous as that of a policeman suddenly becoming a flower child.) Possessing a homespun humility, itself a rarity at the hearings, a practical knowledge of agriculture, and by far the best sense of humor exhibited by any witness, he told his story superbly.

Mr. Yannacone: Would you please tell us what you have been doing professionally since you received your doctorate?

Dr. van den Bosch: I was on the staff of the University of Hawaii Agricultural Experiment Station for two years specializing in a study of the biological control of the oriental fruit fly. . . . Subsequent to that period, I spent 12 years at the University of California at Riverside in the Agricultural Experiment Station specializing in biological control of a variety of insect pests and in the integrated control of pests, particuarly of cotton and alfalfa. Then, in 1963, I transferred to Berkeley and have, since that time, been specializing or working more or less in the areas that I just described. . . .

I teach; I have graduate students; and I do a considerable amount of foreign work seeking parasites and predators of various pest insects. . . .

Q: Now, Professor, in the course of your regular professional activities, do you have occasion to render advice and consultation on the control of insect pests of agricultural crops?

A: Well, my work largely entails research on agricultural pests and recommendations for their control.

Q: And are these recommendations for their control such that they have economic implications? In other words, Doctor, are they put into practice in the actual agricultural industry?

A: Yes, they are.

Van den Bosch then went on to explain that he had been involved with DDT as far back as 1947 when it was being used on alfalfa and observed that, even then, he had seen some secondary problems with it. But then Yannacone guided his witness toward the meat of his testimony.

Q: Now, Doctor, in the regular course of your professional activities, starting with your original association with the University of California, have you had occasion to recommend the use of DDT?

A: Yes, I have.

Q: And as recently as when did you so continue the recommendation of the use of DDT?

A: Well, the cotton recommendations, which were actually formulated last fall, that is, the 1968 cotton recommendations, incorporated the DDT-toxaphene mixture. . . .

Q: Now, Doctor, at the present time, do you still recommend the continued use of DDT in the agro-ecosystem?

A: I personally have decided that I will not endorse this recommendation in the future.

Q: Now, Professor, will you please tell us the basis for this change of opinion?

A: Well, it's the accumulated experience with this material—and here I'm speaking of DDT—its impact on the agro-ecosystem; its direct impact on the arthropod fauna [insects] in the, say, the cotton agro-ecosystem. It's the very recent information that points up the movement of the material out into the biosphere or the world ecosystem, its persistence, and the clear evidence that it is affecting biological mechanisms within the species remote from the area of application. All have contributed to my personal decision that, in essence, there is enough of this material in the environment and there should be a moratorium on its use. . . .

Q: Now, can you tell us why it took so long for you to form the opinion that DDT should no longer be recommended for use within the agro-ecosystem?

A: Well, like many of us, when Pandora's box, so to speak, was ripped open in the middle forties, particularly from an ecological standpoint we were dealing in almost total ignorance. This was the guilt shared by most of the entomologists. There were farsighted people who anticipated these problems, but, nevertheless, most entomologists eagerly seized these [pesticides] and threw them into the agro-ecosystem or into the general ecosystem totally ignorant of the genetic and ecological implications in their use. For years problems developed, such problems as environmental contamination, secondary pest outbreaks, pest resurgence, resistance to insecticides and, either through intellectual ineptitude, ignorance, inertia, or some other factors, these developments simply were not grasped in their total significance.

However, they did set the framework for scattered people over the world to begin to develop studies as to the why, the reasons, the basis for these problems. And, in a very long sense, I simply have to say it's taken time and tremendous effort, tremendous amounts of manpower to develop the background studies to pinpoint these [pesticide] problems so that they will be accepted, as you might say, scientifically valid. . . .

I am a scientist. I work on the basis of the scientific approach, I think that way, and I have to prove things to myself scientifically. I am speaking now as an individual in the evolution of my thoughts on this matter.

Now much of this information is coming together very rapidly. And the picture that has evolved is very distressing to me.

You see, as an experimental or as an applied insect ecologist, I have a very broad range of responsibilities. I have a responsibility as a scientist to develop adequate, sophisticated, high-caliber derived data of this sort. I have the economic consideration of the agricultural community that I serve. I have the responsibility to the general population as far as fitting my findings into the general environment. And all of these things [occur] simultaneously. We have to fit [them] into a pattern which will tell us: Is practice A, which will kill pest A, outweighed by its disadvantages as regards chlorinated hydrocarbons in tuna in the Japan Sea? . . . So it's a very complex thing.

And, frankly speaking as an agro or an applied insect ecologist, we have not received either the administrative or the financial support for the kind of studies we are doing. If there has been concern about the environment and pollution, it has been in the area of public health, in the area of wildlife safety, not in the area of the actual agro-ecosystem itself, that is, what happens to arthropods, the non-target species in the system. So we have been, in effect, working on a shoestring with, in many cases, almost administrative indifference.

[This] is a long way of saying why it has taken us this long a period of time to come to these decisions. . . .

Q: Well, Professor, at this time, in order to form an informed professional opinion as an applied insect ecologist, is it necessary to have information from other diverse disciplines?

A: Yes, it is. In fact, even though some of us aren't aware of this, we are developing what might be called a systems approach to pest control. The inputs, the elements concerned, the diversity of the problem, the complexity of the problem simply means that we can no longer approach pest control either unilaterally as individuals, that is, [through] research, or through unilateral techniques such as to have fully objective chemical control or biological control or cultural control or genetic control. It's a system, we call it integrated control. And all over the world this philosophy, this approach is developing.

It so happens, I suppose, in California . . . we have been the most aggressive about it. Maybe this derives from the fact that we are the greatest users of insecticides in the United States, perhaps the world. We have run the gamut of problems. We have more or less survived our period of penance, and we are now working our way out of the jungle, so to speak, into this more rational approach.

Van den Bosch proceeded to amplify his position on integrated control and its implications:

Mr. Yannacone: Now in the course of your systems approach to pest control and the understanding of pest problems, do you develop personal expert competence in these diverse disciplines, or do you secure such assistance---

Dr. van den Bosch: . . . As I say, our approach is evolving. . . . When we consider pest control in the ecological context, we don't simply have an animal there that we decide must be killed; we have sociological considerations. . . . When I recommend a procedure for pest control, I don't only have the economics of that particular crop and that particular pest, I have 20 million people in the state of California, their water, their air, their soil, their recreation. You might say I have San Francisco Bay and the Pacific Ocean, and perhaps the penguins in Antarctica to bring into this thought.

Now there are areas in there outside of my range of competence, and we have economic considerations. Someone says it's not our business to worry about the economics of pest control, they say all we are supposed to do is kill insects. Well, this of course is a horrendously erroneous idea. But we have to bring in economists to analyze the economics of our programs. This again, is outside the pale of my competence. So we are bringing in economists; we are making economic studies of pesticide use; and we are relating them to the sociological aspects, the ecological aspects.

As you see, it becomes a very complex and very integrated type of situation with a great number of specialties and competences involved.

Insofar as I'm concerned, this is the pattern of modern pest control; this is the way we are going. . . .

Van den Bosch then turned to the important concept of what constitutes a pest.

Dr. van den Bosch: There are two very major considerations in this integrated approach. One is the ecosystem. That includes you and me and the birds and fish and the atmosphere as well as the crop and its pests and its natural enemies. The other is what we call the economic threshold or the economic injury level. And [here] —and we have found it particularly true in the cotton crops—the fundamental practice of the last 20 years [has been] to kill an animal because it was there.

Well, after years now of intensive studies with our two key [cotton] pests, the bollworm and lygus bug, [we find] that the somewhat arbitrary economic thresholds that were employed are totally invalid. In fact, with both species we find, in most instances, more or less regardless of their abundance, that these animals are not causing damage, that is, damage that will compensate for all the other things you do when you --- it is an

artificial control measure. Now we find these creatures doing
damage in certain situations, at certain times, at certain places.
And our responsibility is to identify those situations and attack
[these] creature[s] in those places and at those times. So right now
I would anticipate that we are going to have a very flexible type
of economic threshold, or variable thresholds for these two pests.

Then he went on to describe biological or unilateral control in a
simple way.

Dr. van den Bosch: In its unilateral sense, [biological control] is the
use of parasite predators and pathogens for the regulation of pest
abundance. I guess you could say that that's the classical sense.
It implies manipulation of these things. However, the school that
I belong to, which is the DeBachian school, . . . considers biological controls as a natural phenomenon, essentially the regulation
of, in this case, insect abundance or numbers or populations by
biotic agents.

There is a difference, you see, because—and this has been at
the root of many of the problems that, in the last 20 years [have
occurred]—there is this great natural force, one of the very large
elements of what we call natural control that is comprised of these
parasitic predators and pathogens, these biotic agents that regulate
animal numbers and plant numbers . . .

Because your artificial practices have so often disturbed this
more or less quite unrecognized hidden complex of species, we
have had problems, especially with the use of broad-spectrum,
ecologically unsophisticated insecticides.

Then he went into control.

Dr. van den Bosch: . . . I went into the semantics of this word "control," because it's a very, unfortunately, a very misunderstood
word; to the man who has a cotton field full of bollworms that
are destroying a third of his crop, control is the decimation of
that isolated or localized population. He does this normally by
applying an insecticide.

"Control" in the broad context means regulation which has a
long term and areawide connotation. So we have to clear up the
point here [of] what we mean by "control."

He then described various modes of unilaterally controlling insect
pests.

Dr. van den Bosch: We have genetic control. This can take several
manifestations. The more or less forthright manifestation is the

development of a . . . plant variety resistant to a particular insect. Now that is control in the long term sense. . . .

There is . . . a classic case where they brought the male from two subpopulations of a mosquito—at least on an experimental basis—from one extreme of its distributional range . . . [and have] released it, at least experimentally, in Malaysia [at the other extreme of its range] against females of its own species. The matings produced infertile eggs. This is a long story, but exemplifies genetic control. . . .

Then there's cultural control, the use of agricultural practices, late planting or how you irrigate, how you plant your seed, when you plant your seed. We are developing agricultural control or manipulation of the lygus bug in California through the strip planting of alfalfa in cotton fields. The bug has an affinity for alfalfa, it has an affinity for *leguminous* plants in that group and is actually attracted into these strips and remains there rather than infesting the cotton.

Dr. van den Bosch then told the classic DDT story.

Dr. van den Bosch: . . . The coastal plain of Peru is essentially rainless. However, streams drain out of the Andes through valleys and empty into the Pacific Ocean. There is a sequence of about 40 of these self-contained ecosystems up and down the coast of Peru, each of which is separated by a severe desert. So here we have a little entity, the Canete valley, that has all the elements of, say, the Missouri River valley or San Joaquin valley. . . . It's 50,000 acres of greenery and animals and so on and so forth. Well, they grow cotton in this valley and sugar cane and vegetables.

In the late forties they instituted an [insect] control program based on the then available chlorinated hydrocarbon insecticides for cotton pest control. Within a period of a relatively few number of years following this practice, several of the insect pests had developed resistance to these insecticides; new pests had developed; and, in essence, they had more or less created a monster. With the development of the resistance, they shifted to other kinds of insecticides [in the same chlorinated hydrocarbon family] and the problem worsened to the extent that they had doubled the number of pest species, their pest control cost had spiraled, and their yields were the lowest in a decade.

Then [the farmers] threw up their hands and said: "How do we get out of this?" And they went to an integrated control program and since then they have had their highest yields.

The point I'm making here is here in this case, picking specifically a chemical pest control, but in a very dramatic situation, the unilateral use of one method created a monster, so to speak. And

Pupae of Vidalia beetle with young of cottony cushion scale which they are eating.

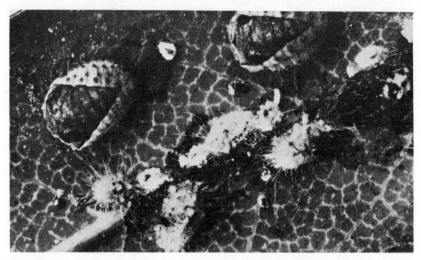

Bollworm larvae go into hibernation after which they will emerge as moths which will be sterilized and released.

If there is a group of animals that has met the competitive challenge of man and held its own, it is the Insecta. Abundance, diversity and adaptability are the key characteristics which have helped insects to stand up to their more clever competitor. And quite ironically, man, the thinking animal, has actually played into his enemy's strength by relying overwhelmingly on a single combat technique—chemical control.
Robert van den Bosch

Bollworm larva 24 hours after exposure to a virus.

it's there in the record for everyone to see; it's a lesson for all of us to learn.

Examiner Van Susteren: May I interrupt for a moment.

What you have just said is you cannot utilize any specific system of pest control, there has to be a generalized attack on the whole problem utilizing the biological, the genetic, the chemical, the natural, and so on, all in a certain balance?

Dr. van den Bosch: Yes. The important point is that we have to smoothly integrate these things.

And more stories about the un-magic of the unilateral use of persistent pesticides flowed from van den Bosch.

Dr. van den Bosch: . . . In the 1940's, when DDT's insecticide properties were discovered, it was like a magic pill had been dropped into the hands of man. Prior to that time he's had to use "lousy" insecticides, some plant roots or leaves, or some heavy-metal-type materials [like arsenic]. They weren't very good insecticides, but they weren't biocides. . . .

You all know about lead arsenate in apples and other things, but [arsenic compounds] didn't tear up the environment. We had no experience with the genetic, the adverse genetic or ecological characteristics of the modern insecticides.

Now the modern insecticides, the synthetic organic materials, were developed by chemists and toxicologists, and they were largely exploited by people who were thinking in terms of the economics; I mean economics from the standpoint of marketing.

No ecological thought whatsoever went into these materials originally. Essentially none is going in now. And this is the basic root of the problem. These are ecologically crude materials, and they have an enormous overall impact on the environment, when we consider there are probably a million and a half species of insects and insect-like creatures. . . .

It was then McLean's turn to work on van den Bosch, but the "scientific convert," as one Madison newspaper called him, proved too tough a customer to be bullied.

Q: Doctor, in your direct testimony you mentioned a problem of lack of sufficient funds for your area of inquiry. Isn't it true that this is a universal complaint we all have?

A: Well, I suppose that's the more or less patented lament of the bureaucrat or the research scientist.

I think perhaps, as you are aware, there's a—has been a—I hate to use the word "attack," but a "complaint" in certain quarters that the universities are shirking their duty to the chemical industry because we aren't screening insecticides any more. . . .

You see, what happened . . . is that we got, I calculated, for the University of California at one time, I think it was, 350 or 400 thousand dollars, something like that, over a certain period to do some kind of research on chemicals.

Now I and the people in my capacity didn't get one dollar. But the research that these funds engendered was screening of insecticides, new insecticides and their economic exploitation. . . .

Q: Economic evaluation?

A: Yes, that's a better word. Exploitation sounds evil.

This, in other words, had a mushrooming effect on the problems which confront ecologists like me. It shoved one set of problems ahead, and it set us back. I spent the last five years, about half of my time, evaluating insecticides. I am a biological control specialist. . . . I evaluated insecticides because of their impact on our [environment]. I want to know what their impact is on the beneficial or the *entomophagous* arthropods in the environment. If the impact of these chemicals is deleterious, when it comes to making the insecticide recommendations for cotton and alfalfa in California, I bring those data to light in these meetings and I say no or I say yes.

But when I work with the natural enemies of the bollworm, say—I have about a dozen and a half to study—that is, little insects which eat the bollworm, parasitize it or sting it . . . (and every candidate chemical that comes into our system for screening more or less is put to the test on the basis of these criteria), I get no external support for this kind of study.

This struck a sensitive nerve for McLean, and he tried to show that what van den Bosch was saying was incorrect; that industry's approach to university screening of pesticides was the right path. But the entomologist was determined to show how the industry's unilateral approach to killing insects was dangerous, in that it took no ecological factors into consideration.

Dr. van den Bosch: This may be the traditional pattern, but . . .—this relates to the registration of insecticides for use by the federal government—they seek certain criteria, mainly efficacy of the material [and] data on performance [and] on the mammalian toxicity. These data are inadequate for our purpose in California, you see. We do not accept federally registered insecticides for use on cotton in California, because there is no local data collected. . . .

There is a tremendous burden on me and my colleagues to obtain these additional data, because there may be 25 materials registered for use on cotton by the federal government [and] a thousand salesmen in the state of California who have every right to recommend these materials. . . .

We have to further screen these materials. It is necessary that we do this, because if we don't get the adverse data, then these materials are going to be sold in California. And many of them have very adverse ecological characterisitics. . . .

McLean saw he was getting in over his head and changed tacks. He questioned van den Bosch about the possible dangers of importing insects into new areas, mentioning a bee which was not noxious in Africa but very much so in South America. He asked:

Q: Do I recall reading recently of an importation of a Brazilian—I mean an African bee into South America that prospered and became quite a problem to people down---

A: That's right. . . .

Q: It was not thought to be a noxious insect in Africa, but it became so in South America?

A: That's right. This is the type of thing we in our organization for the introduction of "beneficial" insects through our system of screening would more or less preclude.

Now I don't know whether the bee was a deliberate introduction into Brazil or whether it was accidental, but it has turned out to be a noxious insect.

I believe the gypsy moth was brought into the United States because somebody thought it was a lovely insect; the Klamath weed was brought into Australia because some sentimental European had an affinity for its lovely yellow flower.

We are talking about two situations. When we import insects
for the suppression of other insects, we are basically dealing with
a phenomenon known as density dependence. We go through
a series of screening processes, [and here] we have 70–80 years
of experience. These are special kinds of insects adapted to prey
upon or parasitize other insects. . . . Because of the density-
dependent nature of these insects, they act as governing agents.
If they are effective, they cause the crash and permanent suppression of the pest they are introduced to control. Being density
dependent, their numbers then crash, also. And thereafter, a state
of regulatory equilibrium is maintained where the pest and parasite more or less fluctuate in an interrelated way at relatively low
density. . . . There is no case on record where one of these things
has become noxious.

At this point McLean tried to bring up the example of the starling
as a bird introduced into this country which has become a serious
pest. Van den Bosch replied:

Dr. van den Bosch: But may I garnish that statement a bit?
So are about half our major pest insects, exotic species. This
is the reason why they are pests, because they have, through one
medium or another, escaped their native habitat and the adapted
parasites and predators that affect them in those habitats; they
have gotten into our environment, and they flourish.
The principle of classic biological control is to attain the
adapted natural enemies of these escapees, bring them back into
association with their host, and hope they will regulate them at
this permanently subeconomic damage level.
So the matter of immigrant exotic species in our North American
environment is very commonplace. In fact, my Dutch father is
one of them.

Further information about integrated control was to come forth
from two other witnesses for the petitioners. The first was perhaps
the world's most eminent specialist in biological control, Paul De
Bach, an entomologist and Professor of Biological Control at the
University of California at Riverside. He, too, testified on the alternative to the heavy broadcasting of chemical pesticides.

Mr. Yannacone: Now, Professor, would you tell us just what is meant
by the term "biological control of insects"?
Dr. De Bach Biological control is the effect of natural enemies in the
regulation of host population densities. In the broad sense, man
doesn't have to enter into the utilization of natural enemies. In
other words, they can bring about biological control on their own,

and of course, frequently do. I might carry it further and say this is the commonplace thing in nature among insects and crops and natural ecosystems to find biological control of various insects and other arthropods occurring.

When you get into crop ecosystems, entomologists frequently speak of biological control from the standpoint of man's activities and what we do to either keep biological control from being disturbed or to augment it.

You can think of three phases of biological control: Importation of natural enemies from abroad, from some other area; bring[ing] in new natural enemies to take care of pests you may have that don't have adequate enemies. Another phase is to conserve the enemies that you have in an area. In other words, get rid of deleterious activities. [These] can be of various types, such as adverse chemicals; you may modify adverse effects of cultural practices such as some types of mowing or harvesting or merely raising dust, which is deleterious in some cases. . . . You can provide requisites that natural enemies need; for example, sources of nectar or honeydew, sources of shelter or places to build nests for predatory wasps.

Then another possibility, which is sometimes called augmentation of natural enemies, involves the mass culture in insectories [and] the periodic colonizing of natural enemies in the field. This is a feasible procedure, for example, when . . . let's say, you have a very effective natural enemy, potentially effective, but the winters are a little too cold and it doesn't quite come through . . . in the numbers that would enable it to successfully control the pest at a very low level. So, by raising these in the insectory and colonizing them in the spring, say getting them started sooner [so that] they would naturally build up later on, you may be able to keep this pest at very low levels throughout the world. This is being put into practice more and more today in many places throughout the world. . . .

Then De Bach spoke of integrated control.

Dr. De Bach: I look on integrated control as an attempt or a means of maximizing the utilization and effectiveness of natural enemies in any crop ecosystem, and doing this by developing the methods which are least injurious to natural [pest] enemies which are always common in any given crop, regardless of where you are. . . . What I am saying is that every crop has many potential pests which can be made accidentally, if you do the wrong things, into serious pests. So in integrated control you utilize the methods which are, say, least injurious to all these potential pests and their natural enemies so that you keep them as minor or non-pests,

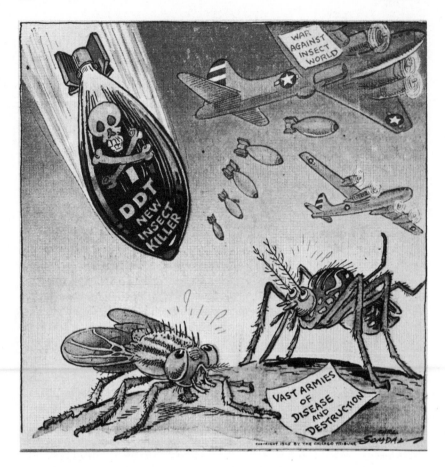

1945

and you utilize a method, then, to treat the one or two or perhaps three pests which [still] occur in a particular crop.

So integrated control can use chemical control in connection with the conservation [and] utilization of natural enemies; it could use cultural control; it could use, for example, sterile male techniques; [or] various modifications and new approaches in entomology that are being studied today and are being tried out. . . .

Mr. Yannacone: Do you want to tell us how chemical pesticides can be used to augment the biological control portion of an integrated control practice?

A: Chemical pesticides come into an integrated control program from the standpoint of being the best materials that you can find, experimentally, to control the pest or pests which don't have . . . adequate natural enemies. In other words, perhaps economic damage will occur if you don't treat these pests.

[Then] your objective is not to find an insecticide which will kill this one pest and, perhaps, say kill 99 per cent of it. [This]

1969

is the old-line thinking in entolmology: . . . you get the poison
that kills this one thing you are thinking about, and you forget
everything else. [Then] you get into the situation where so many
growers and agriculturists around the world are today. [Instead,]
you test various alternatives and then you pick the insecticide for
the pest which you find by experimental study really requires an
insecticide—and this may not always be easy to tell when you
are studying a disturbed ecosystem. There are many pests today
which are secondary, which have good natural enemies, except the
natural enemies are unable to operate due to upsets and disturbances from other practices going on in the agro-ecosystem. So,
by studying these various alternative insecticides, you pick out
the one which gives the best overall control of the whole system.
And this is what you call integrated control.

Then De Bach turned to his own experience with DDT. Beginning
with World War II when the pesticide was being hailed as a "wonder

drug" for agriculture, he showed how, in van den Bosch's words, the chemical had opened a Pandora's box of side effects.

Dr. De Bach: Well, I first used [DDT] when I was in the United States Department of Agriculture in the southern states . . . in 1943–1945. . . . [There] the white-fringed beetle is an extremely difficult thing to control, because the larva is the stage that does the damage, and these are in the soil. So, during the latter part of the war, we got something like 2½ pounds of DDT for testing. (This was one of the high priorities at that time. . . .) It was, well, entomologically, the hottest thing you'd ever heard of. We were able to put [DDT] on a few acres—on very small plots. And it really looked good compared to the material we had before.

From the first mortality counts we made, [it] looked so fantastic that the story went around that we were going to completely eradicate the white-fringed beetle. Of course, this never happened. We still have white-fringed beetles.

[The] idea was prevalent throughout the country at that time [that] DDT was . . . the ultimate solution. (I have seen that [idea] applied to other things as well.) But, as we know now from experience . . . [DDT] wasn't the final answer. . . . To my knowledge we have never achieved eradication with any other insect, that is, as a species, any place in the world. . . .

. . . Upon my return shortly after this work in the south of California, the fame of DDT had spread. It was found to be a rather effective control for the citrus thrips in the Central Valley of California, and it was enthusiastically applied in 1945. In 1946 they found that the cottony cushion scale, which had been the first great example of biological control in the world, had become the major pest in this whole valley. . . .

The cushiony scale, up to then, had been a rare insect since 1890 when it was successfully controlled by a beetle, a predatory beetle called the Vidalia that was brought in from Australia in 1888 and 1889. This beetle was very successful within a year in reducing the cottony cushion scale to such low levels that it was never again of any consequence. In fact, [it was] generally hard to find, until it was upset in 1946 and 1947 by the widespread use of DDT in the Central Valley.

We were called up there. We found that citrus groves were literally encrusted in some cases with the cottony cushion scale to the extent that trees were actually killed.

I don't know how many of you have had experience with how difficult it is to kill a citrus tree. It's not easy because, particularly with any kind of an insect, you can get tremedous infestation [which] may kill twigs or branches. But, to kill an entire tree, it really has to be an enormous infestation. But this is what we had.

I saw entire large groves defoliated by the cottony cushion scale due to the killing of the predacious lady beetles by DDT.

And these groves, for example, didn't come back. The ones that were defoliated might not have been killed, but they would go without crops for perhaps two or three years.

The situation was so serious among the growers up there then, that they were actually trying to buy, and were buying when they could get them, these beetles for one to two dollars apiece.

In the winter of 1946, we surveyed intensely throughout this whole area and we couldn't find a live Vidalia beetle. Now this is over literally hundreds of square miles. In 1947 of course, this infestation [continued]. It was obvious to anyone at this time what the cause and effect were. [But] at first it wasn't, you see, because at that time people didn't have experience with this phenomenon [of resurgence], they really didn't realize that insecticides . . . like DDT could do this.

And their first answer, their first thoughts among the entomologists were: Gosh, it must be something in the climate, something must have changed to do this. But it soon became obvious to everyone that there wasn't any other explanation. We were able to prove, by experimental tests, by testing the beetle, and by comparative tests in which DDT was put onto plots and not put onto other plots that this great upset occurred due to the use of DDT.

So we reintroduced the beetle into the valley in 1947 and modified and, in fact, essentially dropped the use of DDT up there. . . . And since that time, there hasn't been any real problem with cottony cushion scale there or any other place in California. . . .

De Bach furnished a similar story about the California red scale which also became a major pest after the use of DDT, and then he began talking again about integrated control.

Dr. De Bach: . . . We need to point out to growers and entomologists and agriculturists in general that 90 per cent of the potential pests in a given crop have good natural enemies. In order to do this, [we can] upset these natural enemies, in other words, kill them or decimate them to the point where they are no longer effective in controlling the pests. And by comparing this plot where you eliminated them or decreased their effectiveness with an adjacent plot where you didn't do this, you can show whether biological control is, in fact, occurring.

So we found, as I say, early that DDT was a very good candidate material to do this sort of thing; and I have used DDT since . . . to disrupt agro-ecosystems by putting DDT at regular intervals in certain dosages which will kill natural enemies but not kill the pest insect which these natural enemies would normally keep under control.

For example, I could put DDT on citrus trees and kill them by getting rid of the natural enemies of the California red scale or the citrus red mite. The red mite in California is now, and did become very rapidly, . . . a major pest following the first usages of DDT there.

Yannacone then asked De Bach what he thought about DDT's place in integrated control.

Dr. De Bach: No, it's certainly not compatible. It's by far the worst material that I can imagine to try to use in an integrated control system. DDT is so long lasting, so persistent, and so generally deleterious to natural enemies across a broad front . . . that it's, as I say, the worst material I could think of to try to work into an integrated control program.

By the time De Bach testified, Willard Stafford was handling the cross-examination of witnesses for the industry's DDT Task Force. The smooth trial lawyer from Madison almost immediately became involved in an argument with De Bach over the biological control of specific insects in Wisconsin.

Mr. Stafford: Now, is it your testimony that there is a biological control which could be used effectively and economically in the state of Wisconsin for the control of aster yellows disease . . . ?

After an argument over the question, De Bach replied:

Dr. De Bach: I can say that there may be an effective enemy here now. But you couldn't tell if there is if you are using DDT for the control of this leafhopper. . . . I've seen this situation time and time again where DDT is used: a grower is treating for a pest which has effective natural enemies, but which are precluded from action because of this.

Stafford threw out a few more questions at De Bach, but, unable to get through his intellectual defenses, he allowed him to get off the stand unscathed.

The next agricultural specialist to appear was Dr. Alfred Chant of the University of Toronto. He added more data on integrated control.

Dr. Chant: . . . Most organisms, and certainly most insects, the vast majority, 98 or 99 per cent, are being controlled at the present moment without the intervention of man at all. These species are being controlled by factors in their environment, largely parasites, predators, or disease organisms that we have to take no action against at all.

From that point, many entomologists recently have arrived at the conviction that this kind of regulation or control is what we must be striving for with the few species of insects that do create economic problems in our agricultural and forest industries. This can be achieved, and had been achieved in many places . . . by biological control, which has proved to be a success in many cases around the world: [there are] over 200 instances of successful biological control.

The point I want to make here is that the success in biological control is most emphatically and demonstrably a product of the research effort that goes into [it]. The parts of the world, notably California, some European countries, Canada, and Australia, that have invested sufficient time and effort in research in biological control are exactly those countries where this method has overwhelmingly paid off. Where entomologists choose to ignore or neglect the possibilities of biological control and don't support it in the way it should be supported, obviously it doesn't pay off. . . .

The purpose of any pest control program is to protect crops and thereby save money. And it's axiomatic that the cost of the control program must not exceed the value of the crop saved. It is the determination of the balance between the cost of a program in all of its ramifications and the savings to the agricultural or forest industry that determines the so-called economic threshold.

Insect populations vary from time to time, from season to season. And, in every instance, the insect population will vary from below the economic threshold to a point above the economic threshold. When the insect pest is varying upwards and approaching the economic threshold, is the time, and the only time, that one should consider putting into operation some control mechanism to prevent the pest population from exceeding that point where the cost of control is balanced by the cost of saving. . . .

The economic threshold is a very complex statistic. It depends, not only on the obvious things—the cost of the material or the control program being used, the market conditions of the commodity that's under consideration, the cost of labor at the time, and this type of thing—but, over and above that, most people realize there are many other components to be considered. . . . These may have to do with the possibility of creating problems elsewhere from pest control activities, the possibility of adverse effects to other areas, other ecosystems, other factors in the environment than the one [with] which you might be primarily concerned.

Van Susteren then interjected this comment which Chant agreed with.

Examiner Van Susteren: What you have just said, then, is that the same type approach that you use in this thing is the type of approach that an industry might use in regards to cost accounting?

Then Chant continued.

Dr. Chant: One of the chief advantages of thinking in terms of economic thresholds is the realization that must follow from this consideration that one must not use pest control practices unless they are required. That sounds rather naïve and simple, but I think you will find, if you look at local spray calendars all over the United States . . . that, in most instances, this principle is ignored.

There is no point in using a pest control practice unless the pest you are concerned with is approaching or exceeding the economic threshold. And yet most spray calendars don't point this out to the grower, and suggest, in fact, routine applications at so many days interval. In some instances, it's been demonstrated that these control practices are applied even when the pest is totally absent, let alone at a level below the economic threshold. . . .

The present use of pesticides could be reduced by 50 per cent today on the basis of the knowledge that we already possess . . . Two principles would be involved in this. The first principle is: Do not spray until necessary. . . . The second one, and possibly even more important, is not to try to achieve unrealistic levels of control or indeed eradication. What one is simply trying to do is to return the pest population to a level below the economic threshold.

In the long run, the three agricultural witnesses who testified at Madison may have been the most important ones at the hearing. For, despite all of the testimony about the direct environmental damage being caused by DDT, we still must feed a huge population.

This economic necessity presents extremely complex problems needing the sorts of complex answers being provided by such men as van den Bosch, De Bach, and Chant; men experienced in the practical difficulties of intensive agriculture as well as in the problems of preserving the ecosystem on which all agriculture and food production is based.

Yet it is difficult to talk about widespread integrated control of insect pests in a country where the chemical pesticide industry does a billion dollar a year business, and can exert enormous pressures, both on the farmer and the legislature. Nonetheless, if this country doesn't start listening to the words of the integrated control specialists, then its problems will continue to multiply like those insect pests on a cotton field.

10

The Men Who Can Poison the World

The man who is laughing has not been told the terrible news
Bertold Brecht

Harry Hays, as Director of the United States Department of Agriculture's Pesticides Regulation Division, controlled the procedures by which some 13,000 commercially registered products were unleashed on various types of pests and on the world as well. But, during vivid cross-examination by Yannacone—the high mark of his Madison performance—the public learned from Hays the terrible news; those official procedures offered almost no protection at all against toxic pesticides.

Hays's direct examination was conducted by Kenneth Robertson, a U.S. Department of Agriculture attorney, appearing as intervenor in the hearing. Throughout, the testimony was uneventful and bland. Its content seemingly revealed the phlegmatic nature of the Pesticide Division's day by day existence; the functional, if not dull and simple, quality of the processes necessary to protect the public against pesticides.

Mr. Robertson: Dr. Hays, you stated you are the Director of the Pesticides Regulation Division. Will you please state the functions of that division?

A: The function of the Pesticides Regulation Division is to carry out the provisions for registration and enforcement pursuant to the Federal Insecticide, Fungicide, and Rodenticide Act [FIFRA].

Q: Dr. Hays, will you please set out or state the procedures that are followed in connection with registration of an economic poison pursuant to the provisions of the FIFRA?

A: . . . The two primary functions in the Act are registration and enforcement, and we have in the Pesticides Regulation Division a Registration Branch and an Enforcement Branch.

Now under the Registration Branch we have a group of registration specialists as well as a staff of competent scientists in the various areas of disciplines involving pesticides. To register an

economic poison, the applicant must first submit a formal application for the economic poison; he must submit a proposed label,
a statement of the chemical composition of the product, and
effectiveness data and safety data in support of the application.

The application and the data are first reviewed by what we call
the New Chemicals Evaluation Staff. The chemical composition
is reviewed for its accuracy: the ingredients statements, and the
proper nomenclature, the net content, and the product name. The
application, the label, [and] the data are then submitted to what
we call the Product Evaluation Staff. Here we have a group of
entomologists, agronomists, plant pathologists, bacteriologists,
animal biologists.

In general the criteria that are used (that have been submitted
to me by the professional people) include such things as the pest
to be controlled, the dosage and the rate of application, *phytotoxicity*, metabolism, migration, translocation, and persistence.
From this they then review very carefully the directions for use
as proposed by the applicant to see whether or not the product,
used in accordance with the directions of use, would in fact be
effective.

The data and the label and the chemical composition are then
submitted to the Safety Evaluation Staff. Here again the staff is
made up of specialists in biology and toxicology. They review the
data submitted in support of safety insofar as the directions for
use. This would involve, in general, such things as the oral, acute
oral, dermal, and inhalation toxicology; subacute studies designed
primarily to determine if the product has cumulative effects; [and]
subacute feeding studies. Eye and skin irritation is a very important
part in terms of the applicator, [and] such things as sensitization,
reproduction, and carcinogenicity tests.

Now from this it is possible then to determine what signal word
would be used on a pesticide container. There are three principal
words: "Danger," "Warning," and "Caution." In addition, the data
in support of safety would provide a means of determining what
precautionary statements, when complied with, are adequate for
the protection of man and vertebrate animals. . . .

If the product is to be used on food or feed crops and there
is a likelihood that a residue would remain from such use, this
matter is transmitted to the Food and Drug Administration for the
establishment of a tolerance. This usually involves a very extensive
petition, considerable data on feeding, and long-term studies for
evaluation by the scientists in Food and Drug.

When all of the review has been completed by each of the
individual staffs, it is then reviewed in its entirety by representatives of each of these divisions so that we can take a look
at it as a whole. If the data appear to us to be adequate, or to
our scientists, and all provisions of the registration have been met,

the product is registered. If, on the other hand, the data do not appear to be adequate in support of effectiveness or safety, the applicant is so notified of the deficiencies and the need to submit additional data.

In 1963 the President's Science Advisory Committee recommended in its report on the use of pesticides, that other departments in the government should be consulted [so they might] provide information and advice to the Pesticides Regulation Division prior to registering any product. In 1964 there was drawn up what is known as an interdepartmental agreement which was signed by Secretary [of the Interior] Udall, Secretary [of HEW] Celebrezze, and Secretary [of Agriculture] Freeman.

This agreement states that all labels, all proposed labels, applications should be referred to the proper agencies within those departments such as Welfare; to the Food and Drug Administration; to the Bureau of Veterinary Medicine; and to the Department of the Interior for review and advice as to the adequacy of the labeling and the adequacy of the data in support of the registration. This has now been in effect since 1964. Applications are submitted to the interdepartmental groups on a daily basis; we receive their advice; we take their advice seriously in terms of the adequacy of the labeling to protect the interests of the public health aspects and fish and wildlife.

Now the second important activity, of course, in the Federal Insecticide, Fungicide, and Rodenticide Act is the enforcement. And so we have under constant surveillance an inspection system [designed so] that our inspectors, who are located in various geographical areas of the United States, will collect samples of the products that have been shipped in interstate commerce. I stress again interstate commerce.

These samples are submitted to the laboratory for biological testing, and results are sent to the Washington office. If the product is found to be in violation of the Act, is misbranded, it is then subject to criminal prosecution.

For the protection of the public, we have in this past year sampled over 8,000 products that have been shipped in interstate commerce. [They] have been sampled and analyzed by our various laboratories.

In addition to the general functions of the division, we have, of course, a departmental committee on pesticides in the Department of Agriculture. . . . We get information from this committee as well [on] the research activities that go on within the Department of Agriculture. There is [also a] Federal Committee on Pest Control that [has representatives from] many agencies in the government that review[s] the federal programs before any are implemented. We also have a very close working relationship with the state officials under the Association of the American Pesticide

Control Officials, with our primary aim to have as uniform procedures and registration and enforcement as possible. We have in
the past two years met with representatives of the states at some
eight regional meetings to discuss our problems at both the federal
and the state level. We believe this has been a very important part
of our regulatory function.

That, I believe, Mr. Robertson, states our procedure.

Q: Dr. Hays, you set forth the procedure in connection with submission of data, and review of all that data, to support the registration of an economic poison.

Is there any provision for canceling the registration of an economic poison?

A: Yes, Section 4.c. of the Act provides that when the Secretary deems
it necessary to cancel a registration, he shall so notify the registrant, and the reasons therefor. These reasons must be based on
good evidence for such a cancellation. I think the President's
Science Advisory Committee report on the use of pesticides
emphasizes this point, that there needs to be some relief for
industry from any arbitrary or capricious act on the part of a
regulatory agency, and they have a right to file a complaint or
object to any cancellation.

Short, sweet, and placid; like the calm before a hurricane, was the
way someone described Hays's direct testimony.

Cross-examination was a different matter. Yannacone, with his
courtroom instinct for the jugular, had been waiting—so he later
said—to get Hays on the stand for two years. For, according to Yannacone, the further his efforts against DDT evolved, the more apparent
it became to him that the real source of the problem lay not with
the pesticide itself but with its regulation. Yannacone felt that to have
a rational pesticide program which wouldn't either threaten the entire
biosphere or destroy American agriculture, would require a rational
way of regulating pesticide use. And here was Hays, representing the
entire slapdash method of regulating "economic poisons," as pesticides are euphemistically called, sitting in front of him.

Mr. Yannacone: Dr. Hays, how long have you been in charge of the
division?

A: Since July 1, 1966.

Q: What was your job prior to that time?

A: I was with the National Academy of Sciences as a director of the
advisory center of toxicology.

Q: And your Ph.D. was in what, sir?

A: Biology. . . .

Q: Where did you work prior to the National Academy of Sciences?

A: I was formerly at the Wayne State University College of Medicine.

Q: Doctor, the Federal Insecticide, Fungicide, and Rodenticide Act refers in certain areas to provisions of the Food and Drug Act, does it not? . . .

A: The Act itself does not.

Q: But in the Act there is a provision that, under certain circumstances, certain material is referred to the Food and Drug Administration for evaluation?

A: Not in this Act.

Q: Will you tell us how the Food and Drug Administration and the U.S. Department of Agriculture act together in the registration of those pesticides that leave residues on food crops?

A: That is under the Federal Food, Drug, and Cosmetics Act.

Q: And that Act then refers to what?

A: To the requirement of the establishment of a tolerance under the Miller Bill of the Food Additives Amendment.

Q: Now which agency initiates this tolerance procedure for pesticides that leave residues on food crops?

A: The applicant initiates the request for a tolerance if it is to be used on food or feed.

Q: Does the applicant make the initial determination that there will be a residue?

A: Yes.

Q: Is this checked by your department?

A: No, sir. Chemically, you mean?

Q: Yes.

A: No.

Q: In other words, then if the applicant says there is no residue or will be no residue, your department does not check that statement?

A: We look at the data, sir, we review the data submitted with the application to see whether or not there would, in fact, be any residue if the applicant has said there was no residue. . . .

Q: Who supplies the data?

A: The applicant.

Q: From his own research?

A: Yes.

Q: In other words, a chemical company furnishes you data from its own research?

A: That's right.

Q: And if it doesn't measure any residues, you don't check [the] statement that there were no residues found?

A: We do not.

Q: Does anybody?

A: I would imagine that Food and Drug may test the method.

Q: Didn't you just say, Dr. Hays, that Food and Drug doesn't evaluate pesticides unless the petition is brought to them?

A: We are talking about data and residue data.

Q: You have gotten me a little confused, Doctor. I think you testified that the petitioner, the applicant for the registration, submits to your department---

A: Yes.

Q: Certain data?

A: That's correct.

Q: Let's assume that data has an indication that there are no detectable residues on food stuffs---

A: Yes. . . .

Q: The applicant, the registrant, submits the data, right, to your department?

A: That's correct.

Q: And if that applicant says there's no residue detectable on the food-stuffs to which the pesticide [is] going to be applied, your department does not scientifically, analytically check that statement, does it?

A: We check it; not by the laboratory method---

Q: You read his data?

A: That's correct, we read the data.

Q: All right. And you evaluate it; and if it is logically consistent within itself, you accept it, right?

A: That's right.

Q: The only thing you can determine is internal inconsistencies in the data?

A: That's right.

Q: Now at that point if the applicant does not tell you that there is going to be a residue and if his data, internally consistent within itself, shows no residues, there is no referral to FDA under the Food, Drug, and Cosmetics Act, is there?

A: That's right.

Q: Now, Dr. Hays, have you ever seen the registration application of DDT?

A: The original?

Q: Yes.

A: No.

Q: Have there been any subsequent applications for either the registration or reregistration or further consideration of DDT?

A: Yes.

Q: When was the most recent?

A: I have no idea. . . .

Q: Well, Doctor, do you know anything about the registration of DDT?

A: I know there have been a number of registrations for DDT. . . .

Q: What did you say your job title was with the department?

A: I am the Administrator and Director of the division.

Q: Of?

A: Of the Pesticides Regulation Division.

Q: And unfortunately, as bureaucratic operations are conducted, the buck stops in pesticide registration with you?

A: That's right.

Q: You have never seen a DDT formulation registration statement?

A: Oh, yes, I have seen registrations; but I have not actively participated in each registration.

Q: That isn't what I asked you, Doctor. I asked: Did you ever see any?

A: Oh, yes, I have seen---

Q: Okay. When was the most recent you saw?

A: I wouldn't have any idea, sir.

Q: You have been with the department since 1966, right?

A: Yes.

Q: Prior to that time did you examine any?

A: No.

Q: So between 1966 and now in 1969---

You are still with the department, right?

A: Yes.

Q: Still with the same title?

A: Yes.

Q: You have seen some DDT registration statements, have you not?

A: Yes.

Q: Now in those DDT registration statements was data furnished as to any sublethal effects of DDT?

A: If they were, they were submitted to the Safety Evaluation Staff.

Q: You don't know?

A: I wouldn't know.

Q: Doctor, who asked you to come here and testify?

A: The Department of Agriculture wishes to discuss the procedures used in the registration of pesticides.

Mr. Robertson: Mr. Examiner, the Department of Agriculture filed a petition for leave to intervene in this proceeding as a result of learning of the proceeding and analysis of the record disclosing that the federal registration procedures were discussed.

I don't think that Dr. Hays personally is in a position to say who may have requested him. This determination was made within the department.

Mr. Yannacone: All right, that's good enough. Dr. Hays, you did know why you were coming here?

A: Oh, yes.

Q: You did know the purpose of this hearing?

A: Yes.

Q: You are a representative of the U.S. Department of Agriculture?

A: Yes.

Q: Now, Doctor, what was your official job title again?

A: I am the Director of the Pesticides Regulation Division.

Q: And you are in charge of the regulation of pesticides, right?

A: Yes.

Q: And you are responsible for the regulation of pesticides?

A: I am responsible to see that the activities of the registration are carried out by those assigned to the duty of reviewing each application.

Q: Okay. And the scope of your duties or the extent of your duties is defined in the Federal Insecticide, Fungicide, and Rodenticide Act officially?

A: Right, yes. . . .

Q: Do your duties comprehend a study of the safety of DDT and its metabolites as they may be formulated as economic poisons?

A: Not directly.

Q: Is there any other department in the federal government that you know of that is responsible for the approval of the registration of an economic poison for use in interstate applications, other than the Department of Agriculture?

A: There's no other department responsible except the Department of Agriculture.

Q: Your department, the United States Department of Agriculture, is wholly and completely responsible, then, for determining whether or not an economic poison may be used in interstate commerce, right?

A: That's right.

Q: And is there any other division within the Department of Agriculture other than the one that you are the head of that is responsible for the approval of a particular registration for use?

A: There is no other division.

Q: In other words, then you are the top of that division of the U.S. Department of Agriculture which is responsible for determining whether or not a particular economic poison, in this case say DDT, is registered for use in interstate applications, right?

A: That's right.

Q: All right. Now, Doctor, tell us from your duties and the duties of your division as set forth in the Act as you read it and it's interpreted to you by your department's legal talent, tell us, Doctor, what specific information about a pesticide being proposed for registration your department is interested in.

A: I leave that entirely, sir, to the people responsible for the various scientific disciplines within the division.

Q: All right. Doctor, there is a policy and there are rules and regulations set forth as to what information a registration application must contain, is that right?

A: In general, yes, sir.

Q: What are these general requirements, please?

A: I think I have stated those.

Q: We want to review them for the record.

A: Chemical composition---

Q: All right, could we stop for a moment?

A: Yes.

Q: This is simply the chemical formulation of the compound as it's going to be used, right?

A: Right.

Q: And I think you testified that this is checked by your staff for accuracy?

A: That's correct.

Q: It's also checked for nomenclature in that it conforms with whatever the current scientific nomenclature for the substance is?

A: Yes.

Q: And if it's a substance like technical DDT, which is a mixture of *isomers*, your department checks to make sure that the isomer mixture concentration is set forth on the label accurately, right?

A: Not necessarily.

Q: All right. Isn't that part of the chemical composition?

A: If it's a technical grade, it need not state on the label what the percentages are.

Q: But does it have to say what the isomers are?

A: No.

I just gently stood there and sought the truth!
Victor Yannacone

Q: In other words, then a technical grade such as tech DDT need only state on the label that its major constituent is thus and so?

A: That's correct.

Q: And that's all your department then checks for, right?

A: Yes. Checks.

Q: Okay. Now what's the next element that's checked for?

A: The next would be the matter of effectiveness.

Q: Now, effectiveness. Will you tell us what to your agency "effectiveness" means?

A: I again, sir, rely entirely upon the scientists within the division of that discipline to determine what they consider to be effective. . . .

Q: With respect to DDT, . . . the check would be by entomologists?

A: That's correct.

Q: And they would be checking on effectiveness, right?

A: Yes.

Q: The effectiveness they check for is what?

A: Whether it controls the pest.

Q: The target insect?

A: That's correct.

Q: Now when we say "pest" in your department, we are referring to "pest" as defined by the Act, are we not?

A: That's correct.

Q: All right. Are we referring to any kind of insect that isn't defined by the Act?

A: Not that I would know of.

Q: In other words, then, a pest is like an officer and a gentleman; it's determined by an Act of Congress and set forth in the Act; and if it's named as a pest in the Act, it's subject to the jurisdiction of your department? . . .

Examiner Van Susteren: Just a moment.

Mr. Robertson: The term "pest" is not used in the Federal Insecticide, Fungicide, and Rodenticide Act. The term "insect" is used as well as "fungus," and so forth; so just for clarification I thought I would---

Mr. Yannacone: All right, we will take counsel's advice. Dr. Hays, where is the word "pest" defined then if it's not in the Federal Insecticide, Fungicide, and Rodenticide Act?

A: I don't know of any other place anywhere it's defined.

Q: It's defined in Title 7 of the U.S. Code, isn't it, in the Agriculture section?

A: As we just stated, it did not mention just "pest" but specified "insects."

Q: It sets forth in that section that these insects are subject to control, doesn't it?

A: Yes, I guess you would say that.

Q: For the purpose of policy determinations at your level in the department, effectiveness is considered as what? In other words, what do you understand by "effectiveness"?

A: As I said, I'm relying solely on the scientists to determine what, in their opinion, would be an effective control.

Q: This effectiveness then, is determined by an entomologist on a staff, right?

A: That's correct.

Q: Aren't there any published guidelines as to what is effective or not effective control?

A: No published guidelines.

Q: Aren't there any internal memoranda or understandings at various levels of your department that might tell us what "effective" is?

A: I'm sure that this could be found in many scientific journals.

Q: Oh, Doctor, you are the head of the only section of the USDA that is responsible for the registration, which means the actual interstate sale ultimately, of economic poisons like DDT. And you have told us that one of the criteria for registering these economic poisons is their effectiveness. You're telling me that effectiveness is left to the independent judgment of some technician or some entomologist on a scientific staff well down the line, so far down the line in your department that you don't know what his criteria for effectiveness are?

A: We have chief staff officers in the Product Evaluation Staff who are not well down the line, but who are competent entomologists, agronomists, plant pathologists.

Q: All right Doctor---

A: They are not technicians.

Q: It's a matter of opinion. . . . Now the safety data that's submitted with the registration statement, who furnishes this in the first case?

A: The applicant.

Q: All right. And are there set forms or set criteria or set elements of this safety type data? In other words, what do they check for and furnish you in the way of information?

A: Well, Mr. Yannacone, the first information that is provided by the applicant is data on the acute oral, dermal, and inhalation toxicology.

Q: All right, stop for a moment so we can expand on that. Does this include $LD_{50's}$?*

A: That is correct. . . .

Q: What kind of data comes in in addition to $LD_{50's}$?

A: It includes cumulative studies, repeated studies, repeated daily doses; it includes, as I said, eye and skin irritation studies---

---

*LD_{50} means the lethal dose of a pesticide necessary to kill 50% of a group of test animals, usually laboratory rats. (Eds.)

Q: These are on animals?

A: That's correct.

Q: What else is included in this initial safety evaluation?

A: In addition, there are studies on reproduction.

Q: All right. Now when you say studies on reproduction, what are the usual kind of reproduction studies that are considered in that?

A: In laboratory animals.

Q: And what do they measure, fecundity or fertility?

A: That's correct.

Q: They measure basically the number of offspring reproduced and whether there's any statistical difference between the control and the sample?

A: That's correct.

Q: Anything else?

A: Sensitization.

Q: And this again is with an experimental animal population?

A: That's correct. And at times human patch tests are involved.

Q: All right. And this is by normal allergic reaction study procedure?

A: That's correct.

Q: Anything else included?

A: In some instances antidote studies.

Q: All right. This is assuming that there is a poison problem, they determine the basic antidotes, and they furnish you with that information?

A: That is correct.

Q: All right. And that would then be included on the label?

A: That's correct.

Q: And by the way, Doctor, so you understand where we are going, I'm following the same outline you did on your direct testimony, which was nice and complete.

Now, [the evaluation of data] is done by which staff? . . .

A: This is done by the Safety Evaluation Staff, by review of the interagency staff of the Public Health Service.

Q: All right. Now the New Chemicals Evaluation Staff does what?

A: They are primarily concerned with the ingredient statement on the label, the net content, the product name, any matter dealing with flammability, the nomenclature of the compound, whether or not it is consistent with the chemical abstract nomenclature.

Q: Now, the Product Evaluation Staff does what?

A: Well now, let me try to make something clear here, Mr. Yannacone, in that we have two general groups of compounds that we would categorize as non-food-use compounds and those which may be used on food or feed. So that if we have an insecticide that does not in any way have any connection with food or feed, then our Product Evaluation Staff in the insecticide section will review the data in terms of whether the product is effective against that particular insect which is named on the label.

Q: I see. Now in other words, any substance such as DDT might have multiple registrations; and if it was going to be used for the control of certain insect vectors such as mosquitoes, and not on human foodstuffs it would then be evaluated as a non-foodstuff pesticide?

A: That is correct.

Q: And if it were going to be used for a particular food crop, it would then be evaluated by a different group within the section or under different criteria?

A: It would be reviewed by the same entomologists, but in addition would be reviewed by the Food and Drug Administration.

Q: I see. All right, now the review by the entomology group for an insecticide that is only to be used on non-foodstuffs includes only a study of its effectiveness against the target organism, is that correct?

A: That's part of it.

Q: All right, what else does it do?

A: Then we get into the question of where it might be used. [We get] into areas involving fish and wildlife.

Q: All right. Now would you elaborate for us on how this work with fish and wildlife is done?

A: Well, there are certain data that are required by our division in support of the registration, in . . . any use that might possibly affect our fish and wildlife, and studies would be required, or data would be required to see what doses would, in fact, affect any fish or wildlife.

Q: All right. Would you back up a moment? Who makes the determination on whether or not there will be an effect on fish and wildlife?

A: The applicant usually is quite cognizant of the need for any other studies as it might pertain to fish and wildlife.

Q: Now what kind of studies are presented to you generally on data with respect to fish and wildlife safety?

A: Well in general, as I recall the scientists' review, they require LC_{50}* concentrations in a variety of fish species; they require LD_{50} studies for certain types of birds.

Q: All right. Anything else?

A: I can't recall, at the moment, anything else.

Q: Do you recall whether they do reproduction studies with fish?

A: We have not, in the past, required these extensive studies on reproduction. But this again would be in concert with the Department of Interior in terms of advice from them as to what would be needed.

Q: All right. Now when you speak in terms of advice from the Interior

---

*LC_{50} is the lethal concentration in water. (Eds.)

department, is this advice in any way binding on the Department of Agriculture?

A: Well, I don't know that any advice is always binding. We certainly do take into consideration any information and advice that the Interior or any other agency would give us.

Q: Are you familiar with the testimony of Dr. Lucille Stickel, head of the Pesticide Research Group at Patuxent for the U.S. Department of the Interior, which was made a part of this record; and [of] Dr. Kenneth Macek who is with the Fisheries Section at Columbia, Missouri, on the effects of DDT on certain birds and fish, respectively?

A: I am in general familiar with it, yes.

Q: And is there any doubt on the part of your department experts as to the validity of this data, that you know of?

A: I wouldn't know what their feelings are in the matter at all.

Q: Well, have you, at the policy level, considered this data?

A: I have not thoroughly reviewed all of the data; [I am] just in general familiar.

Q: I want you to assume, Doctor, then, the substance of the data and the testimony of Dr. Stickel and Dr. Macek, Dr. Stickel having testified that she observed in laboratory populations eggshell thinning and reproduction failure in kestrels, a bird of prey, and ducks, a particular kind of duck; and Dr. Macek observed reproduction failure in the lake trout with sublethal concentrations of DDT at levels now already present in the respective environments that these species inhabit naturally.

Now assuming that, Doctor, is there anything you in your capacity can or would do about the registration of DDT?

A: Well, we would have to have, certainly, some very extensive and definitive data as it pertains to the normal usage of any pesticide, and not based solely on any laboratory finding.

Q: Well, Doctor, do you require this type of data from the applicant when he makes up his registration statement?

A: We have not required this kind of data in the past; although we have recently reviewed our criteria on our data for fish and wildlife and have indeed added other kinds of data such as field studies that we think will be very useful. But it is too early now to evaluate this kind of approach.

Q: . . . When did you make the changes?

A: Oh, in about the last year we have been requiring field studies particularly in areas where there's very large and heavy wildlife populations.

Q: Does this apply to a chemical that is already registered such as DDT?

A: Yes.

Q: And who performs these experiments?

A: We have requested the applicants to consider and to initiate studies in the matter of field testing.

Q: All right. And what kind of field tests are you comprehending within this kind of study?

A: Well, we have contemplated putting certain types of bird species such as pheasants and ducks in areas where we know that this could well be a problem and to see whether, from the normal use of the pesticide, there is, in fact, any serious hazard associated with such use.

Q: What about the fact that a great many of these experiments have already been done, both by private individuals working for academic institutions and by the U.S. Department of the Interior itself?

A: We consider the data from a variety of sources, not only what we get from the applicant, but what is available from whatever source.

Q: Let's back off a moment, Doctor.

You are now aware, you have testified, of the work of the U.S. Department of the Interior, Dr. Macek's fish work, and Dr. Stickel's work with the hawks and the ducks, right?

A: Yes.

Q: I take it from this that you don't question their scientific accuracy. And you have considered them laboratory studies. You should by now also be aware of a great many field studies that have been done over the past five or six years.

Now, Doctor, isn't this more evidence against the use of DDT than was ever submitted on the safety of DDT to your department originally on the registration of DDT?

A: Well, there's no doubt but what there has been new data presented in the last few years on a variety of pesticides about which we knew little a few years ago.

Q: All right. Doctor, whenever possible at this hearing we are trying to discuss DDT and its metabolites exclusively. I know this might restrict you a little bit, since you are responsible for a great many more pesticides than just DDT. However, DDT is the subject matter of this action.

Now isn't it a fact, Doctor, that there is already accumulated in the scientific literature and in the data submitted to your department from the U.S. Department of the Interior, more data indicating damage to wildlife populations on a broad scale from DDT than there is data on the safety to wildlife populations already in your records? Isn't that a fact, Doctor?

A: I wouldn't go so far as to say that is a fact.

Q: All right. Doctor, then you tell us for the record what data your department does have on the safety of DDT to wildlife populations that conflicts with the data that you already are aware of, [indicating] that DDT is not safe to wildlife populations.

A: Well, I think we can say in general that, from the wide usage of DDT, there has been a remarkably good record of the use of DDT without any evidence of any widespread adverse effects. . . .

Q: Doctor Hays, have you ever heard of Dr. Robert Risebrough?

A: Yes.

Q: Are you aware of any of his published material?

A: I am aware that he has published material.

Q: Have you evaluated any of that published material?

A: I have not.

Q: Are you aware of the general subject area of that published material?

A: In a very general sense.

Q: What is your understanding of the subject material?

A: It, in general, is a review of the widespread effects on wildlife.

Q: Of?

A: Of DDT.

Q: Have you ever heard of Dr. Charles F. Wurster, Jr.?

A: Only during his testimony.

Q: Then you are not aware of any of his publications?

A: No, sir.

Q: Do you read *Science?*

A: I read a lot of journals, not *Science* particularly. . . .

Q: What journals do you read, Doctor, in the regular course of your daily duties?

A: Well, I read the *Journal of Toxicology and Applied Pharmacology,* the *Journal of Pharmacology and Experimental Therapeutics;* articles that are brought to my attention by our staff, the journals of which I am not particularly aware of at the moment, however.

Q: Basically then you read the two big toxicology journals?

A: That's correct, yes. . . .

Q: Is the subject of the status of DDT and its metabolites as an economic poison under consideration at this time or under investigation at this time by your department?

A: The subject of DDT is under consideration by the Department of Agriculture through the National Academy of Sciences, National Research Council.

Q: Is this the only pesticide that's being considered in this way?

A: No, it involves all persistent pesticides.

Q: I see. But what is the interest of the Department of Agriculture, the division that you head that's responsible for registration, in DDT and its metabolites, if any?

A: I just remarked that we are interested in the total concept of persistent pesticides including DDT as it relates to our environment, and have asked the National Academy of Sciences to make such a study.

Q: But what about your department, Doctor; what work is your department doing, if any, on this matter?

Examiner Van Susteren: On what matter?

Mr. Yannacone: On the matter of DDT.

Examiner Van Susteren: Well, is it evaluation, or re-evaluation?

Mr. Yannacone: I don't know. I would like to know from Dr. Hays what, if anything, other than referring the matter to the National Academy of Sciences, has his department done with respect to DDT (which apparently was [a problem] of sufficient magnitude . . . to refer to the National Academy of Sciences)? . . .

Dr. Hays: We are not doing anything.

Mr. Yannacone: Okay. Now in other words, then, the registration of DDT as it now exists is not under direct review by your department now?

A: No, sir. . . .

Q: But now, Doctor, you testified on direct examination that among the criteria used by the Product Evaluation Staff are evidence of phytotoxicity and metabolism of the pesticide, migration of the pesticide, translocation of the pesticide, and persistence of the pesticide; is that correct?

A: That's correct.

Q: Okay. This data then is furnished to you by the applicant, is that correct?

A: That's correct.

Q: And it's not checked independently in a scientific analytical sense by your department, is it?

A: That's correct.

Q: In other words, Doctor, your department is dependent for data in these registration cases on the applicant's good faith, isn't it?

A: The Act requires the applicant to submit the data.

Q: Yes, but your department doesn't independently verify or check this data?

A: No.

Q: Okay. Now you probably don't have the money, do you?

A: That's right.

Q: But you do have a staff of analytical chemists and biologists and what not that do work and just check this data, don't you?

A: Check only from the point of view of enforcement.

Q: I see. And the enforcement provisions of the Federal Insecticide, Fungicide, and Rodenticide Act are solely limited to whether or not the product is mislabeled?

A: That's correct.

Q: And the label need only contain its proper chemical name and ingredients and contents and weight and a general safety warning?

A: And directions for use.

Q: Directions for use. All right.

Now what data is generally required for a pesticide such as DDT on the matter of persistence? Soil persistence?

A: When DDT was introduced, I---that's been many years ago---

*In effect, then, federal registration requires no testing of
the impact of insecticides on the insect community to
which they are applied, or their potential for triggering
pest resurgence and secondary pest outbreaks.*
Robert van den Bosch

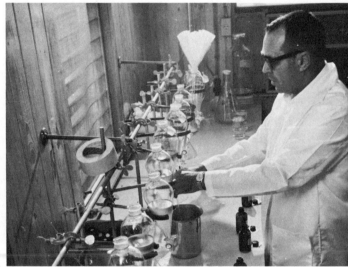

*To test for pesticide residues in soil, samples are col-
lected, isolated in distilled water . . .*

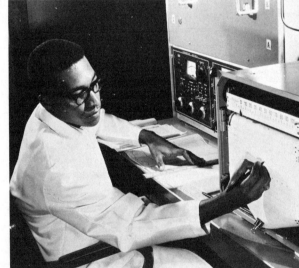

*. . . mixed with chemicals which absorb any residues,
and analyzed in a gas chromatograph.*

Q: No, I'm talking about these current new registrations. You said since '66 there have been some registrations of specific uses of DDT. What kind of data---

A: Persistence in soil.

Q: That's all?

A: Essentially.

Q: I don't want to ask you a technical question unless you feel you want to answer it. But if you can tell us, tell us how this soil persistence data is gathered; in other words, what kind of tests do they make; do they just measure the amount of DDT applied, and then the amount in the soil later on?

A: I would have to leave that detail to the scientists.

Q: Okay. But the only persistence data you are interested in in the department as far as registration is concerned is soil persistence?

A: I said mostly in the soil, in general. I'm sure that water is also included.

Q: Well, persistence in water. Do you have any idea how they measure persistence in water?

A: I'm sure that one could take samples of water at varying times to see how quickly it is either hydrolyzed or is broken down in the presence of water. If it's highly soluble in water, it may persist for some time.

Q: Okay. Doctor, we are now talking about DDT. Do you know anything about the behavior of DDT in water?

A: It's very insoluble.

Q: Then how are you supposed to measure its persistence in water? Do you have any idea of its solubility?

A: No.

Q: Would you believe it if I told you that, according to Mr. McLean, who I think is probably accurate, its solubility is something on the order of one part per billion in water? And you know of course, I'm sure, from your toxicological studies that the relative minimal sensitivity, accurate sensitivity in the gas chromatograph is on the one part per billion range. So we are talking about a substance which is soluble in water only at the lower threshold of detectability.

Now you don't really measure honestly and accurately persistence in water, do you, for DDT?

A: Not for DDT.

Q: All right. Now you indicated that translocation is a consideration?

A: That's right.

Q: Now again speaking with reference to general criteria, what kind of data do you get on translocation of pesticides?

A: Well, we are, I'm sure, primarily concerned with the uptake of a pesticide from the soil to the plant and whether it's translocated from the roots into the plant.

Q: Any other types of translocation?

A: I don't know.

Q: All right. And this translocation data would be with reference to the plant on which the pesticide, in this case DDT, is applied?

A: That's right.

Q: All right. You are not interested in, generally, translocation from, in, through, and about plants that might be non-target plants?

A: No.

Q: In fact, most of this information and research is directed towards the target insect and the target plant, is that correct?

A: That's right. . . .

Q: Now, Doctor, there are no studies that you know of submitted with these registration statements on the effects of DDT on phytoplankton, are there?

A: I'm not in that field at all.

Q: I think you also indicated that studies are made of the migration of the pesticide, is that correct?

A: That's right. . . .

Q: . . . Your department in reviewing the registration of this pesticide is only interested in its migration through soil?

A: No, I'm sure they are interested in many other things. I wouldn't know particularly just what their scope of interest would be. . . .

Examiner Van Susteren: By "they," you mean the evaluators in your division or the evaluators of the applicant?

Dr. Hays: The evaluators of the applicant as well as the review scientists. . . .

Mr. Yannacone: In other words then, the evaluators in your division essentially cooperate rather extensively with the applicant in determining criteria for measurement, don't they?

Dr. Hays: Yes, I would say we work very closely together in developing good criteria.

Q: Good to or for whom, Doctor?

A: For the---for everyone concerned.

Q: All right, Doctor. In other words, then, the real mission of your department is the greatest good for the greatest number; isn't that so?

A: I think that's right, yes.

Q: And you sincerely believe that your department is now operating for the greatest good for the greatest number, don't you?

Examiner Van Susteren: Well, Counsel, the Examiner will break in and say that you are using a very good Marxian doctrine when you start using the phrase, the greatest good for the greatest number. And where are we going here?

Mr. Yannacone: I don't particularly like to be referred to as a Marxist. I think we can point out some differences of approach between the local Marxists and yours truly. But I will---including no beard---but I will---

Mr. Robertson: Along those lines, Mr. Examiner, I also would like to inject that as Dr. Hays has testified in his direct testimony, the mission of this particular division within the Department of Agriculture is to see to it that the provisions of this Act are complied with in all cases.

Mr. Yannacone: Okay. That's what I wanted to get on the record.

Now, Dr. Hays, your mission in your division in your department is to see to it that the Federal Insecticide, Fungicide, and Rodenticide Act is met in all its requirements? Isn't that right?

A: That's correct.

Q: Of your own knowledge of the Federal Insecticide, Fungicide, and Rodenticide Act, can you point to any portion therein where considerations are given to, for, or about fish and wildlife? . . .

Examiner Van Susteren: Dr. Hays has already stated on direct that there is an interdepartmental agreement between Agriculture, Interior, and Health, Education and Welfare.

Mr. Yannacone: I'm aware of the agreement, Mr. Examiner, but I want the statutory provision in from this witness that governs his department.

Examiner Van Susteren: If he knows it, he can tell us. If he doesn't---

Mr. Robertson: Well, can I---

One of the provisions in the statutes is that a product will be misbranded "if the labeling accompanying it does not contain directions for use which are necessary and, if complied with, adequate for the protection of the public";---

At this point I would inject that this term "public" has always been considered to mean man, all of his beneficial animals, his wildlife, his livestock, and so forth.

Mr. Yannacone: All right.

Mr. Robertson: Now the second provision deals with the warning or caution statement. A product would be misbranded "if the label does not contain a warning or caution statement which may be necessary and, if complied with, adequate to prevent injury to living man and other vertebrate animals, vegetation, and useful invertebrate animals. . . ."

Mr. Yannacone: And now, Doctor, you are aware of those two provisions?

A: Yes.

Q: Can you tell us now whether or not any new data has been required by your division for DDT since 1966?

A: Not to my knowledge.

Q: Doctor, you are aware that there is now some evidence that there is injury to vertebrate animals and useful invertebrates attributable to DDT and its metabolites?

A: Yes.

Q: You are aware, are you not, Doctor, that there is at present some evidence that a great many of the world's (in general) and the United States' (in particular) ecosystems are contaminated with DDT and its metabolites; are you not?

A: Yes.

Q: You are aware that DDT is still recommended for use by the Department of Agriculture; are you not? . . .

A: Yes.

Q: But the pesticide is still registered and still may be sold with the label that was on it prior to 1966, is that right?

A: That's correct.

Q: Now, Doctor, is there a procedure available to you or your division upon receipt of evidence that a particular registered compound may be causing some damage, significant damage to vertebrate animals and useful invertebrates; is there any procedure for your department, your division of the Department of Agriculture to take some action?

A: The division could well receive such data if it were available and presented to the division---

Q: Stop just a moment. . . .

Examiner Van Susteren: The division could receive such data and what?

Dr. Hays: For review.

Mr. Yannacone: All right. You have received such data, have you not?

Dr. Hays: Formally? No.

Q: You have not received such data formally?

A: For review. . . .

Q: Doctor, tell us what your department considers a formal request for review?

A: We, as I said, have reviewed data wherever it may be found for our daily responsibilities of the registration of pesticides.

   Now in regard to DDT, we have received no specific request to review any data gathered by anyone for our evaluation.

Mr. Yannacone: Would you please, Madam Reporter, read back my original question. And, Doctor, would you answer that question? . . . What is a formal request for review? How do you make a formal request for review?

A: Well, I don't know there is any formal procedure, Mr. Yannacone. But it would seem to me that if anyone wished to submit any data relevant to the effects of DDT, it can be done by simply sending it to the division for review and evaluation.

Q: All right, Doctor. I don't mean to be nasty. But you just told us your department has not received a formal request for review. Yet you said just before, that you have knowledge of Dr. Stickel's data, Dr. Macek's data, both of which [come] from competent federal government agencies, plus the fact that you know about Dr. Welch's and Dr. Risebrough's data. What is necessary to make a formal request for review? Isn't that enough? Isn't that enough to raise a question in your mind, Doctor? . . .

Examiner Van Susteren: The Examiner has spent all of his, almost all of his adult life in the state bureaucracy. And some of the material and information that is submitted to the various departments of this state in his opinion is nonsense. It would seem that perhaps what Dr. Hays is talking about, that if it comes from what appears to be a responsible source on a subject of some importance with some data that appears to have some significance or validity on its face---I presume that that is the type of situation that Dr. Hays is referring to. . . . Am I correct, Doctor?

Dr. Hays: That's correct.

Mr. Yannacone: Mr. Examiner, you have apparently survived your years with the state bureaucracy with some measure of talent, competency, and public spirit left. What I'm trying to establish now very simply---and I ask the question again: What constitutes a formal request or a request for review that you would consider formally made to your department?

Mr. Robertson: Mr. Examiner, I would like to know what Mr. Yannacone means by a formal request for review.

Mr. Yannacone: This witness stated that he received no formal request for review of any of this so-called new data since 1966 with respect to DDT. I asked him, since he's already testified he has knowledge of this data, personal knowledge, much less certainly the lower levels of his department have knowledge, but he's got personal knowledge; I would like to know for the record and for our own personal information so we can see to it that a formal request is made, what constitutes a formal request.

Examiner Van Susteren: All right, now---

Mr. Yannacone: And he just testified there is no such thing, there is no such procedure.

You can't have your cake and eat it too, Counselor. . . .

Examiner Van Susteren: Is there any provision in the Act which permits the department to act on its own motion in this respect for revocation, cancellation, and so on?

Mr. Robertson: The way the statute is phrased, it is only the Secretary, in accordance with the procedures specified in the Act, that can initiate the cancellation of registration. The Secretary granted the registration, and pursuant to the procedures can initiate procedures to cancel that registration.

Examiner Van Susteren: And the initiation, then, would have to be made by a formal request to the Secretary. Is that your interpretation of the Act?

Mr. Robertson: Right. The Secretary, or Dr. Hays through his division.

Examiner Van Susteren: But it would have to be addressed to the Secretary, as such, for a formal initiation of a re-evaluation or cancellation?

Mr. Robertson: Such a formal request, perhaps, as Mr. Yannacone has been referring to, I think, would properly be addressed to the Secretary.

Mr. Yannacone: And this is basically the procedure under the Federal Insecticide, Fungicide, and Rodenticide Act for cancellation?

Mr. Robertson: That's the statutory authority, yes.

Mr. Yannacone: Okay. To your knowledge, Dr. Hays—or if Mr. Robertson can fill it in—is there any other procedure for the deregistration or cancellation or suspension of a registration that you know of?

A: I don't know of any other procedure. . . .

Q: Doctor, in your operation as chief of your division within the Department of Agriculture, do you observe specifically in your activities any regulations which might be set forth in [the] memorandum which you introduced as exhibit No. 115 [Department of Agriculture, Agricultural Research Service, *Safe Use of Pesticides*, a memorandum between the Secretaries of HEW, the Interior, and Agriculture]? . . . What I'm interested in finding out is: Does your division within the Department of Agriculture have anything to do with anything that might be set forth in that memorandum of understanding? If so, tell us what it is?

A: Well, this is a memorandum of understanding in terms of review of the applicant's product in terms of the label and data for review and consultation with the other agencies.

Q: I see. Okay. It's a review and consultation understanding?

A: Yes.

Q: Now you are familiar with it?

A: Yes. . . .

Q: In the regular course of business of your division within the Department of Agriculture, what specific activities do you undertake to: ". . . keep each of the other departments [being Interior, and Health, Education, and Welfare] fully informed of developments in knowledge on this subject [the subject of pesticides] from research or other sources which may come into its possession"?*

A: We exchange information that we have on any work that may be done that relates to registration of pesticides.

Q: I see. Have you exchanged information with Interior and Health, Education, and Welfare on DDT?

A: I wouldn't know whether information has been sent back and forth or not. This would have been done by the scientists within each of the sections. . . .

Q: Dr. Hays, in the regular course of your work for the United States Department of Agriculture as chief of the division that you are director of, do you have care, custody and/or control of the actual filed documents, the registration applications?

Mr. Stafford: Haven't we had a ruling on this line of questioning? . . .

Mr. Yannacone: I'm asking him whether or not he's got them. They cannot be found. They have been denied to us; they have been denied to Senator Nelson. Now somebody is here from the USDA; let's find out who's got the papers.

---

*Bracketed statements are Mr. Yannacone's.

Mr. Stafford: I ask that the record---that your Honor direct that that statement of Counsel, this gratuitous statement be stricken from the record and expunged from this proceeding.

Examiner Van Susteren: It may stand. But the Examiner has already ruled, Mr. Yannacone, that that request will need to go to Mr. Robertson; he represents the department here today; and as to---

Mr. Yannacone: What is Dr. Hays?

Examiner Van Susteren: No, Dr. Hays does not represent the United States Department of Agriculture here today; it is represented by an attorney.

Mr. Yannacone, are you going to tell me that your various witnesses, or former witnesses here represent the Environmental Defense Fund? . . .

You may submit your request to Mr. Robertson.

Mr. Yannacone: All right. . . .

Now, Doctor, since 1966 have you ever had occasion to examine any of the data or information prepared by any of the departments under your control, such as the Product Evaluation Staff or the Safety Evaluation Staff or the New Chemicals Evaluation Staff, with respect to DDT registration applications?

A: I have not personally reviewed---

Q: Who in your department, if you know, has?

A: I stated before that the scientists within each of the sections are responsible for the review of the data in support of the registration.

Q: All right. Dr. Hays, are you the head of this division?

A: Yes.

Q: Who do you report to at the next higher level?

A: I report to the Deputy Administrator.

Q: Of?

A: Agriculture and Research Service.

Q: What is his name?

A: Dr. Frank Mulhern.

Q: Spell it, please, for the record.

A: M-u-l-h-e-r-n.

Q: And do you report to anyone else?

A: Not directly.

Q: Who reports to you in your division? Directly?

A: The assistant directors.

Q: How many are there; what are their names?

A: Two. The Assistant Director for Registration, Mr. Harold Alford; Assistant Director for Enforcement, Mr. Lowell Miller.

Q: Anyone else?

A: No.

Q: Have either of these two men ever reported to you on DDT or its metabolites?

A: They may have discussed some registration. I would not know what in particular.

Q: Is it in the regular course of your activities as Director of this division a practice to require written reports from either of these two assistants?

A: No.

Q: In other words, then, you conduct all your business with your two associates by conference and by verbal communication?

A: That's correct. . . .

Q: Do you put any directives in writing to either of those two assistants?

A: Yes.

Q: Okay. What kind of directives, Doctor?

A: They are directives that have to do with division activities.

Q: Such as?

A: Procedures.

Q: Such as procedures for registration, or consideration of registrations?

A: Procedures largely in the conduct of our daily affairs.

Q: Which include registrations?

A: It may.

Q: What do you mean, "it may," Doctor? Does it or does it not?

Examiner Van Susteren: Just a moment, Counsel. First of all, perhaps we are all assuming something here that we are wrongly assuming---

Mr. Yannacone: That the witness knows anything?

Mr. Robertson: Mr. Examiner---

Mr. Stafford: I object.

Mr. Yannacone: I will withdraw the comment.

Examiner Van Susteren: It's not only going to be withdrawn, but I think you owe the witness an apology.

Mr. Yannacone: I will owe the witness an apology when I see that he does know something.

Examiner Van Susteren: I feel that a remark like that addressed to Dr. Hays in this type situation is reprehensible.

Mr. Yannacone: All right. I will apologize for the dignity of the Court and the dignity of the profession.

I'm sorry, Dr. Hays.

Examiner Van Susteren: Perhaps we are erroneously assuming that you are aware of some of the intricacies of bureaucracy. But sometimes there are two different lines of authority in an agency. There may be administrative authority and there might be what you might call professional or line authority.

Mr. Yannacone: Okay, let's find out.

Examiner Van Susteren: And perhaps---

I don't know how the Department of Agriculture is set up, but I do know how certain other agencies are set up.

Mr. Yannacone: All right. It can't be set up as badly as some.

[Doctor,] is there a division of functions within your department between professional activities, which involve scientific evaluation of the pesticide, and nonprofessional administrative and clerical activities?

A: Yes.

Q: Now are you responsible for both of those divisions of function?

A: As head of the division, yes.

Q: Now who is directly on a day to day basis responsible for the professional activities or the scientific activities?

A: As I have stated . . . the chief staff officers of each of the sections. . . .

Q: Now who do these chiefs report to? Directly?

A: For registration?

Q: Yes?

A: To the Assistant Director for Registration.

Q: And that man reports directly to you?

A: Yes. . . .

Q: Okay. Since 1966, Doctor—and from now on you can preface everything with that—since 1966 have any reports been made in writing from these section chiefs to your assistant in charge of registration with respect to DDT registrations?

A: I wouldn't know precisely if there had been specific reports that would necessarily have come to my attention.

Q: Do you ever confer directly with these section chiefs?

A: We have a staff meeting.

Q: How often.

A: Usually once a month.

Q: Since 1966 have you at these meetings ever discussed the registration of DDT?

A: That I couldn't remember. . . .

Q: Doctor, since you took over as chief, have you communicated directly, or indirectly through your assistant in charge of registration to these professional section chiefs with respect to the registration of pesticides?

A: Yes.

Q: Have these communications been in writing or verbally?

A: Some have been in writing.

Q: Have any of these communications involved registration procedures applicable to the registration of DDT?

A: I couldn't answer that. . . .

Q: Doctor, did you prepare an outline of your direct testimony before you testified here today?

A: I---

Mr. Stafford: Object to that, your Honor.

Mr. Yannacone: I want to find out what he does know, if anything.

Mr. Stafford: Also on the basis of relevance.

Mr. Yannacone: He testified now "I don't know," "I don't know" about the day to day operations of his department. . . .

Now all I'm trying to find out is whether or not there's any flow of information or communication or regulation or rule or internal policy that might be applicable to the registration or have any bearing on the registration of DDT that flows, not from the bottom up where Dr. Hays might not understand or know or might have been stopped at his assistants' level; but from Dr. Hays down. That's the question I have asked. Has he made any recommendations orally or in writing that might be applicable to the registrations of DDT as they come in from his office either directly to these section chiefs or through his administrative assistants to these section chiefs. That's all I'd like to know.

Examiner Van Susteren: Can you answer the question?

Dr. Hays: No, I don't really; no, I can't remember the details of every little memorandum; I do not know of any. . . .

Mr. Yannacone: Now to your knowledge, since 1966, have there been any changes in the criteria for registration of pesticides, in particular DDT? . . .

Dr. Hays: Your Honor, I think there has been only one change that was recommended by the Chief Staff Officer for the limitation of the use of DDT for cockroaches. . . .

Q: And when he made that recommendation, how did that become enforceable? . . .

A: This would not have required an enforcement action, but rather a registration action.

Q: Would you tell us, please, so we get the record a little bit clearer, what do you mean by registration action?

A: There are times when there needs to be a modification in the directions for use. And where in this instance there was some evidence to indicate the resistance on the part of the cockroaches to the actions of DDT, it seemed that there was no longer any need for this particular use; and therefore the entomology chief staff officer instructed his staff that in the future these uses would be phased out in the case of cockroaches.

Then the cross-examination of Hays went into the kinds of toxicity data required to register pesticides.

Mr. Yannacone: Is there a written standard for evaluation [of data necessary for registration] which governs the activities of each of [the] department section heads?

Dr. Hays: Only in so far as the, let's say the type of statement that would go on the label, as I said, in terms of a signal word. Now this is spelled out.

Q: Now the words we are talking about, if I remember correctly, are "Danger," "Warning," and "Caution"?

A: Yes.

Q: [The] three groups of senior staff professionals are responsible for the determination of which of those three words go on?

A: The Safety Evaluation Staff would be responsible for which word would be applicable in each instance.

Q: I see. Now the safety evaluation chief, who is Mr. Shaughnessy---

A: Mr. McClain.

Q: Mr. McClain. He accumulates and evaluates all the data that's presented on safety and then decides whether the label is going to have "Danger," "Warning," or "Caution"?

A: That's correct.

Q: Is there any review of his evaluation?

A: No, other than his own, or the people working with him.

Q: But there's no higher authority than he on this subject?

A: No.

Q: Now and you take his recommendation without any further work?

A: Yes.

Q: Now what are the basic meanings of those three words?

A: Well, in the case of a product requiring the signal word "Danger," it is based on the acute LD_{50}.

Q: Would you explain that a little bit more in detail?

A: This is the dose that is found or has been found to produce a 50 per cent mortality in the population study.

Q: All right.

A: Now---

Q: Excuse me just a moment. I don't want to interrupt your trend of thought, but on "Danger" the determination is made on acute LD_{50}'s. In what kind of experimental situation or what level of LD_{50}'s?

A: Well, any product that has an LD_{50} lying between 0 and 50 milligrams per kilogram would require that signal word "Danger," poison, skull and cross bones.

Q: Okay. LD_{50} in what?

A: In rats.

Q: Is the experimental procedure to be followed by whoever is submitting this data---And by the way, this data is submitted by the chemical company or manufacturer, right?

A: That's correct.

Q: And it's not checked independently by your department?

A: No, sir.

Q: Now is there any specified procedure for making these tests?

A: No specified procedure. It is a well-recognized procedure that is used by most laboratories.

Q: All right. In other words then, there is nothing in your department at this particular level that's equivalent to a mil spec for testing a particular product?

A: No.

Q: Now the next label down, I assume, the next level after "Danger" is "Warning," right?

A: "Warning."

Q: Okay. Now what's that based on?

A: Any product that has an LD_{50} lying between 50 and 500 milligrams per kilogram would require the signal word "Warning."

Q: And "Caution," which is the next lower one?

A: 500 to 5,000.

Q: All right. And if it takes more than 5,000, you don't put anything on it?

A: It may not necessarily require a caution statement.

Q: All right. Now in other words then, Doctor, the function of the Safety Evaluation Staff is to evaluate the data submitted by the manufacturer with respect to $LD_{50's}$ in rats; and . . . the LD_{50} [level], if it's in one of these three ranges, determines which of the three warning labels goes on it?

A: That's correct.

Q: The product---

Mr. Robertson: Mr. Examiner, this line of questioning, this testimony by Mr. Yannacone with respect to what Dr. Hays has said, the impression to me is being given here that this is the only thing the evaluation staff does. And I don't want that impression. . . .

Mr. Yannacone: No, I don't mean to give that impression.

Now, Doctor, would you tell us what else the safety evaluation group, led by this Mr. McClain does? . . . I want [it] understood, Doctor, throughout, that wherever possible we don't want the broad spectrum, we just want to know what is applicable to a material like DDT.

A: A material like DDT would require, as I said before, the dermal toxicity, the inhalation toxicity---

Q: And again---

A: ---and any other such requirement as the scientists would think would be important in terms of precautionary statements.

Q: But as for fixed procedures in your department, the mandatory procedures, it involves a determination of $LD_{50's}$ in rats?

A: Only for the signal word.

Q: Now are there any other studies done to your knowledge or required by this section other than dermal and inhalation and sensitivity studies?

A: Not other than I have already indicated that the review staff would think would be essential for a proper evaluation for preparing a precautionary statement.

Q: All right. Now these precaution statements that you are referring to, none of us is familiar with the labeling of these products. What's a precautionary statement? What do you mean by that?

A: I'm thinking of a formulation containing DDT that might be highly irritating if it were spilled into the eye.

Q: In other words, the statement would say: "Warning (Caution, Danger), harmful if spilled into the eye"?

A: Or "Avoid Contact with the Eyes."

Q: All right. And is this based on studies again with rats?

A: Rabbits.

Q: Rabbits. Okay. Now these dermal and inhalation studies, the determination is again acute $LD_{50's}$?

B. C.   by Johnny hart

A: Acute and subacute.

Q: All right. What is considered subacute in this sense?

A: 21 day dermal application, 90 day feeding study.

Q: Producing $LD_{50's}$?

A: No, to determine any other effects that would not be discernible.

Q: Such as?

A: Such as cumulative effects.

Q: Such as?

A: Changes in the *hematopoietic* system.

Q: What kind of tests are generally required, Doctor?

A: There again I have said that I have relied solely on the toxicologist to determine what particular test he would require in any given formulation. . . .

Q: Doctor, are you aware from your research and your own personal work as to the types of studies used to determine various kinds of subacute toxicities?

A: I'm aware of them.

Q: Are you familiar with the type of testing required to determine whether or not there is sublethal or subacute damage to the hematopoietic system in an animal?

A: Yes.

Q: Do you know whether or not all the necessary tests for such damage are done with respect to the registration of the pesticide DDT since 1966?

A: I'm aware of a great many studies that have been done.

Q: With respect to the registration?

A: I'm not familiar with all of the data that has been submitted over the period of 20 years.

Q: No, Doctor, that is not the question I asked you; I asked you since 1966, in your department, are you aware of whether or not studies with respect to subacute or sublethal effects of DDT in animals were submitted?

A: No, there have been none since 1966, to my knowledge.

Q: There have been no studies to your knowledge submitted?

A: That's right.

Q: With respect to registration?

A: Not since '66.

Q: All right. Do you have any knowledge of whether or not there were any subacute studies prior to '66?

A: I would have no direct information as to what was submitted prior to '66.

Q: Doctor, again all these studies that are included in the safety evaluation section's consideration are furnished by the manufacturer or the applicant, is that correct?

A: Yes, sir.

Examiner Van Susteren: Except that I believe, Dr. Hays, you stated that the head of the safety section could ask for further and additional tests to be run either by the applicant or by an independent laboratory---

Dr. Hays: Yes.

Examiner Van Susteren: ---selected by the applicant?

Dr. Hays: That's correct.

Mr. Yannacone: Of your own knowledge do you know whether or not the head of the safety evaluation section was called for such additional studies?

Dr. Hays: Since 1966?

Q: Since 1966?

A: I don't know that he has, sir.

Q: Have you?

A: Well, let me say, Mr. Yannacone, there have been no new registrations that would have required the submission of any studies such as that, because most of the registration actions regarding DDT have been on what we call amendments to the registration and not new registrations.

Q: Fine. In other words then, the original, old, sometimes 20-year-old registration applications are simply amended to add or delete a new use, is that correct?

A: If it was a new use, it would have required additional data.

Q: In what type of circumstances would an amendment not require new data?

A: If it was for a use on which we had data that did not extend any additional exposure either to the applicant or to the consumer. . . .

Q: I see. Then in other words, if the DDT were approved for Dutch elm disease control 18 years ago and that the only consideration today with respect to further use in shade tree protection would involve no new data, because it's just another tree?

A: If it is exactly the same dosage, the same rate of application, the same formulation, and all we have added is one additional use, there would be no need for any extensive toxicological data.

Q: In other words, if the original application of DDT, approved in 1947 or 8, for forest pest control were two pounds per acre applied by hydraulic spray, let's say, or by airplane, and if today an applicator wanted to apply 1.9 pounds per acre to a different kind of tree, he would have to submit no further data?

A: I didn't say that.

Q: Well, explain to us in that sense, please, Doctor.

A: A gap of some 20 years would make quite a difference. If a product were registered a week ago and the applicator wanted to extend its use to another area, we would require no additional data over and above what he's just submitted. But we certainly do not look upon data submitted 20 years ago as adequate for present day needs. . . .

Q: Now when is the most recent data, new data, that's been required by your department or division on DDT that you know of?

A: We do not necessarily, Mr. Yannacone, require the applicant to supply additional data if, in the interim, data has already been published in the literature or where other sources have made available such information as we may have required in its absence.

Q: Doctor, I don't mean to pin you down; but I'm still trying to find out what if any data with reference to DDT is relied upon by your department to continue this registration?

A: We ask the applicant to supply that information which we do not have at this moment to support that registration.

Q: Dr. Hays, you have said that during the past two years amendments to DDT registrations have been permitted without new data being required; is that correct?

A: That is permissible.

Q: All right. Well, I don't know whether it's permissible, and I don't care whether it's permissible. I want to know if it has been done?

A: Yes.

Q: You are chief of the division that's responsible for this, right?

A: That's right.

Q: The technical responsibility is one level lower than your administrative assistant, is that correct?

A: Yes, sir.

Q: You confer with these individuals, staff member chiefs on these applications, is that correct?

Mr. Stafford: Your Honor, I object to this line of questioning, it is repetitious. We have gone over this hierarchy three or four times. We have a limited time of this witness, and I ask Counsel to direct his attention to new matters.

Mr. Yannacone: I'm trying---

Examiner Van Susteren: Well, perhaps the witness could answer the $64 question: Has any new specific information been submitted and received by Dr. Hays's division in regards to a DDT registration? Am I correct?

Mr. Yannacone: That's very close to the $64 question.

Mr. Stafford: Well, let's have an answer.

Dr. Hays: I know of no new data that has been submitted other than what we have available from the current literature on which we base a lot of the evaluation. . . .

Mr. Yannacone: Do you require as chief of this division, through your technical people, any evidence that a proposed use of DDT will not result in *translocation* of the material to plants other than the target plant?

A: It's usually directed to the target plant.

Q: That wasn't the question I asked you, Doctor. Answer the question I asked, if you can, or say you can't answer---

A: The target plant.

Q: Does your division require any data as to the migration or mobility of DDT after application other than through soils?

A: No.

Q: Does your division require any evidence of the sublethal effects of DDT or its metabolites on fish prior to registration or as a condition to registration?

A: We have required such data.

Q: What type and kind of data?

A: That has to be determined by the people responsible for this area.

Q: Who furnishes the data?

A: The applicant.

Q: It is not checked by your department?

A: It is not required to be checked by the department.

Q: And it isn't? . . .

A: The Act does not require our testing.

Q: And therefore you don't do it?

A: No.

Examiner Van Susteren: However, you did testify that in this type of a situation, it is given to the Department of the Interior?

Dr. Hays: That's correct, sir. . . .

Mr. Yannacone: Does your division have to accept the findings of the Interior Department? . . .

A: We seek the advice of the Department of the Interior for comments on the adequacy of the labeling and the data in support of the label.

Q: Doctor, adequacy of the label meaning, in other words, if it says on the label "Dangerous to Fish and Wildlife," is that sufficient?

A: This is where we seek the advice of the Department of the Interior.

Q: All right. And if the Department of the Interior says, yes, sir, indeed it is dangerous to fish and wildlife, are you satisfied that the product is registrable if it contains on the label "Dangerous to Fish and Wildlife"?

A: If that is what would be required, yes. . . .

Q: In other words then, Doctor, as long as the product's label says that it is in fact dangerous to fish and wildlife, the requirements of the Act and the duties of your division are satisfied, right?

A: I think we have one point to make, Mr. Yannacone---the words "Danger," "Warning," and "Caution" are used in terms of the human hazards.

Q: Right.

A: There has been no such terminology for fish and wildlife.

Q: All right. Would you stop a moment. Let's backtrack on that. In other words then, "Danger," "Warning," and "Caution" are not measures of damage to wildlife?

A: That's correct.

Q: They are measures of possible damage to humans based upon acute LD_{50} doses in rats plus some other indeterminate amount of information which may or may not be required, is that right?

A: That's right.

Q: Okay. Now with respect to vertebrate animals and beneficial invertebrates, what if any requirements are there?

A: There are precautionary statements on the label regarding the use of the material, "Keep out of ponds or streams" if it involves a hazard to the fish and wildlife.

Q: All right.

Examiner Van Susteren: Just a moment. Could the Examiner interrupt. It would seem to me there is an area of confusion here, and the area of confusion seems to be the labeling requirements and procedures and the registration requirements and procedures; and somehow or another it looks like these things are being used interchangeably and concurrently.

Mr. Yannacone: All right. . . . Is the labeling procedure part of the registration procedure?

A: Yes, sir.

Q: It is one single registration procedure, the basic end of which is a proper label?

A: Yes.

Q: All right. In other words then, Doctor, the requirements of the Act as far as your division is concerned are satisfied when you have got a proper label?

A: That's right. . . .

Q: If the label carries, or the manufacturer agrees to put on his label "Warning" or "Precaution, don't use near ponds and streams" and what not, then the requirements of the Act are satisfied to the extent that your division will now approve this registration?

A: That's correct.

Q: No further consideration then is given by your administration to fish and wildlife once this precaution is accepted?

A: Other than the review by the Department of the Interior. . . .

Q: Are you permitted to register a product which is apparently safe for humans but which is totally damaging to fish and wildlife, let's say has an acute mortality at very low levels; are you permitted to register such a product with a precautionary label?

Mr. Robertson: Mr. Examiner, I believe this testimony has been gone over before to the effect that [Dr. Hays's division] receives the information from the Department of the Interior, . . . and this information is applied in connection with the registration and in connection with the enforcement or compliance with the provisions of this law.

Examiner Van Susteren: So it's quite obvious then after Interior's comments are received, it goes to your safety evaluation, and safety evaluation determines then whether the Interior's comments and guidelines are going to be accepted or not?

Dr. Hays: That's correct. . . .

Mr. Yannacone: This is not done then at your level, at the highest level; this is done down at a low level?

A: That's right.

Q: Now, Dr. Hays, I will rephrase that prior question. I apparently didn't make it clear to Counsel or anyone else. Can you refuse to register a pesticide solely on the grounds that it causes damage to non-target vertebrates?

A: This would be based on the intended use. We do not anticipate that the intended use would result in any such damage, and this is the reason for the precautionary statement by the Interior often requested "Do not use in areas where fish may be present."

Q: But, Dr. Hays, didn't you just testify that the only evidence of translocation you require or that you know of that is required is within the target plant, and the only evidence of migration is through soil; wasn't that your testimony?

A: That's in general, yes.

Q: What do you do with a substance that has mobility by mechanisms other than translocation through the target plant or migration through local application in the soil?

A: What do you mean, what do we do?

Q: Let's assume the Interior department advises you that a particular substance, in this case DDT, is co-distilled with water from a given application and can be transmitted miles and miles and miles away with water vapor?

Stafford now interjected, trying to help assemble the fragmented Hays. But Hays only succeeded in burying further the Pesticide Research Division by admitting that only two products had been cancelled in the past five years.

Mr. Stafford: Now referring to Section 4.c. of your enabling Act, which is the section for cancellation, have you had occasion in the past, Doctor, to cancel registered pesticides under this section? And please exclude from your consideration the many which were cancelled due to the change of nonresident policy. Other than those?

Dr. Hays: There have been two such actions taken since 1964; the first being the cancellation of all registered use of thallium. This was a highly toxic material used in the home and was responsible for a significant number of deaths among children. On this basis the department decided that it was not in the public interest to continue the registration of a product that accounted for such a large number of fatalities.

Q: And there was another product also? Give the name.

A: There has more recently been a proposed rule-making regarding the registration of phosphorus paste. Again, this is a product that was registered many years ago, [that] has been responsible for a significant number of deaths. And in view of other materials of a less hazardous nature, the department decided to cancel these registrations.

Q: Now in view of your experience with prior cancellations and your knowledge of the Act, is it your opinion that the present law affords an adequate remedy to protect the public against registered pesticides which allegedly turn out to be harmful?

A: Yes, sir.

Then Yannacone heatedly began cross-examining Hays again.

Q: Now, Dr. Hays, let's tell this court right now what procedure is available should your department fail to heed whatever the decision is of the National Academy of Sciences, National Research Council, should they recommend the deregistration or the cancellation of the registration of DDT. What procedures are available other than the procedure in the statute that you set forth and was read by your counsel which provides for a proceeding initiated by the Secretary of Agriculture himself? Tell us what other procedures.

Mr. Stafford: Object to the question, because it has already been asked and answered.

Mr. Yannacone: Nonsense! All morning we have heard him tell us he doesn't know anything. And now suddenly he tells you?

Mr. Stafford: Counsel, you don't need to shout.

Examiner Van Susteren: Just a moment.
    Procedure for what, for cancellation?

Mr. Yannacone: Cancelling that registration.

Examiner Van Susteren: All right. We'll go through it again, but---

Mr. Yannacone: No, other than the statute, Mr. Examiner, Section 4.c.

Examiner Van Susteren: We can only assume, Counsel, that if the National Academy of Sciences and so on, and the committee come up with a recommendation that it be banned, the only thing they could do would be to let the Secretary know under the Act, and the Secretary institutes the proceeding.

Mr. Yannacone: Is that a fair statement, Dr. Hays?

Dr. Hays: That's right.

Q: That's a fair and accurate statement of the procedure?

A: That's right. . . .

Q: Has your Safety Evaluation Staff given you any report on DDT since you have been in that department?

A: No.

Q: I see. May I ask who made the request of the National Academy of Sciences [for the DDT study]?

A: As I read, the United States Department of Agriculture. . . .

Q: You did participate in the deliberations that led to the departmental request [to study DDT]?

A: Yes.

Q: Did you review the material that formed the basis of this request?

A: Not all of it.

Q: Did you review any of it, Doctor?

A: Yes.

Q: All right. What kind of material did you personally review?

A: Data that was available in our files, data that was available in the literature.

Q: Such as, Doctor?

A: Some of the reports that you referred to.

Q: All right. Now you made an evaluation of this data personally, right?

A: No.

Q: You reviewed it, didn't you, Doctor?

A: Yes.

Q: Did you review it personally?

A: Yes.

Q: In what capacity? The stuff you reviewed personally---

Mr. Stafford: Mr. Examiner---

Examiner Van Susteren: Just a minute. You are badgering the witness. Give him a chance to answer. . . .

Mr. Yannacone: Doctor, look, you reviewed some of it, you said, in person. Okay.

Dr. Hays: Yes.

Q: I am only interested now in questioning you about what you reviewed personally. You said you reviewed some of the papers we have discussed, Dr. Stickel's and Dr. Macek's, right?

A: That's right.

Q: Those reports among others formed the basis of this recommendation, which you supported, which the department made to NAS through contract, right?

A: Yes.

Q: Now, Doctor, you reviewed this material. Did you evaluate it?

A: I evaluated only to the extent of my concern.

Q: What kind of concern, Doctor?

A: That adverse effects had been reported.

Q: All right. Now, Doctor, were these reports that you had part of any formal request to your department for action from the United States Department of the Interior?

A: The request---

Mr. Robertson: If you understand the question.

Mr. Yannacone: I will rephrase the question.

Doctor, you have stated for three hours that nobody's ever made a formal request to your department for review of the registration of DDT, is that correct?

A: That's correct.

Q: All right. . . . Let's go back a minute, Doctor. Your division is the only one responsible for the registration of pesticides, right?

A: Yes. . . .

Q: No other division in USDA?

A: That's correct.

Mr. Stafford: All been asked and answered, and I have objected to it many times.

Examiner Van Susteren: First of all, we have to recognize—and I am not impugning any Secretary of Agriculture—but involved in all of this, if a formal request came in to the Secretary of Agriculture and he did not pass it on down to Dr. Hays, or it got lost in the maze of bureaucracy, Dr. Hays would have no idea about it, and---

Mr. Yannacone: I know that. I am not binding him with what the Secretary wants to know. All I want to know is what he knows in his division.

Examiner Van Susteren: And he is telling you, Counsel.

Mr. Yannacone: Okay. Let me---

Mr. Robertson: He has told him, Mr. Examiner.

Mr. Yannacone: Well, that may very well be, but he apparently changes his mind and his recollection pretty conveniently.

Mr. Stafford: I object and ask it be stricken.

Mr. Yannacone: That's another question.

Examiner Van Susteren: That's another aside, Mr. Yannacone, and so on. I warned both Counsel yesterday that I was not going to tolerate these insinuations that are of a jeering, sneering nature so far as a witness is concerned. This borders on badgering. And while he is well represented by Counsel here today, the Examiner can only assume the responsibility to prevent any badgering.

Mr. Yannacone: I don't want to badger Dr. Hays. All I want is to get a relatively clear record of a very complicated system. This is pretty obvious from the past three hours.

Now you in your division are solely responsible for pesticide registration, right; this division?

A: That's correct, the division, yes.

Q: Now your department, the United States Department of Agriculture, which is chaired by a Secretary of Agriculture, who is a political appointee and who filters his information or whatever down to you through some other kind of channel, has made a formal request coupled with a financial grant to the National Academy of Sciences, National Research Council, for a certain review of pesticides, correct?

A: Yes.

Q: Okay. Now among the pesticides reviewed are persistent pesticides, right?

A: That's correct.

Q: One of those persistent pesticides is DDT, right?

A: That's correct.

Q: All right. Now I asked you before, and you said you did participate in the departmental level group that supported this application for a review, right?

A: That's correct.

Q: Okay. And you did testify that you supported this application because of concern that you had as a result of some studies which you had personally seen, is that correct?

A: That's right.

Q: This included Dr. Stickel's and Dr. Macek's studies. Right?

A: That's correct.

Q: Did it include anybody else's study that you recall, like Dr. Risebrough?

A: I don't recall what other studies it involved, your Honor. [It involved] a lot of discussions and a lot of our experiences with use of persistent pesticides as they pertain to food and feed crops, the problems that have arisen in relationship to the tolerances established for persistent pesticides, and in particular DDT. There was a whole area of problems that the Department of Agriculture concerned itself with, and therefore asked that we submit a request to the National Academy of Sciences to look at this problem as a whole, not just one isolated problem.

Q: Fine. Now, Doctor, your division is the only division of the Department of Agriculture that has direct responsibility for pesticide registration, is that correct?

A: I think I have answered that before, sir.

Q: All right, and a number of times, and the answer is yes.

Now, Doctor, the Academy of Sciences, National Research Council, is going to render a report, is that correct?

A: That's correct.

Q: That report is going to be to the United States Department of Agriculture, is that correct?

A: That's correct.

Q: Now I think you testified that if that report is accepted by the department, then some action may be taken, is that correct?

A: That's correct.

Q: All right. Now other than this request to the National Academy of Sciences, National Research Council, has your division to your knowledge requested any reviews of information concerning the persistent pesticides such as DDT during the two years you have been there from any agency, including the U.S. Department of the Interior?

A: No, sir.

Yannacone, Van Susteren, Hays, Robertson, Stafford, and the packed hearing room then decended into a maze of bureaucratise that left everyone dazed. Arguments raged back and forth over who was on first, and if whoever was on first knew what the person on second was doing. This comedy of bureaucratic errors, with Van Susteren as umpire, centered around the forms of questions, and levels of responsibility, and succeeded, perhaps, only in mirroring the decay of the English language as a functional mode of communication among government agencies. After verbiage had filled the air for some ten pages of transcript things got rolling again.

Mr. Yannacone: Now, Dr. Hays, you testified that there were two cancellations of registration of pesticides since 1964, thallium and phosphorus paste; right?

Dr. Hays: That's right.

Q: All right. And you testified that these were because of their obvious human toxicity. All right? Now, Dr. Hays, will you tell us whether or not these cancellations were effected in accordance with the procedures described in Section 4.c. of the Act that you operate under, the Federal Insecticide, Fungicide, and Rodenticide Act?

A: I think they were.

Q: And in other words, this proceeding was initiated by the Secretary of Agriculture, right?

A: That's correct.

Q: All right. Now do you know of any procedure whereby a private citizen or a group of citizens can initiate a proceeding for cancellation of registration?

Mr. Stafford: That has been asked and answered.

Examiner Van Susteren: We have gone through this, Counsel, so many times, and we aren't going to hear any more. The procedure is outlined in the Act.

Mr. Yannacone: And that's all? There's no way around it?

Examiner Van Susteren: That's what the witness has testified about ten times.

Mr. Yannacone: All right. Now you were on this committee, Doctor, that made the application through the Department of Agriculture itself to the National Academy of Sciences, the National Research Council, to conduct this 18-month study on pesticide residues. . . . Did this committee issue a report to the Secretary of Agriculture?

A: I think we did.

Q: Did you sign that report as one of the participants?

A: I did.

Q: Do you have a copy of that report?

A: I'm sure I must. . . .

Q: Doctor, was any other technical or scientific information submitted to the Secretary, if you know, than your committee report?

A: I don't know of any technical or scientific data that was submitted.

Q: Okay. In other words then, your committee report furnished the technical and scientific information, right?

A: Formed the basis of our concern and the request that a study be made. . . .

Q: Dr. Hayes, in the regular course of your professional activities have you had occasion to examine any scientific data or studies with respect to toxic effects, lethal and/or sublethal, of DDT and its metabolites on non-target organisms?

Mr. Stafford: Objected to as repetitious.

Mr. Yannacone: I haven't gotten an answer to that question in four hours.

Mr. Stafford: Probably not going to get one.
     Excuse me.

Mr. Yannacone: Then if he can't answer, let him say so.

Examiner Van Susteren: Just a moment, Counsel, just a moment. This refers to Dr. Hays professionally and not necessarily as the administrator of the division?

Mr. Yannacone: Right.

Examiner Van Susteren: And he is a professional witness.
     Can you answer the question? Have you ever done any work in that respect?

Dr. Hays: Research?

Mr. Yannacone: No, I didn't ask research, I asked---

Dr. Hays: Your Honor, I think I have stated a number of times that I have reviewed documents, published and unpublished, that relate to a variety of pesticides, including DDT.

Q: All right. Now, Doctor, did you concur in the request of this committee to the Secretary to engage the National Academy of Sciences and the National Research Council to evaluate pesticides in accordance with this contract?

A: Your Honor, I stated that I signed the report.

Q: Well, does that mean you concurred in it?

A: I must have.

Q: Is this report a public document?

Examiner Van Susteren: It is now, Counsel.

Mr. Yannacone: Could you produce it for this hearing; if you went back to Washington, would you be willing to produce a copy of it?

Examiner Van Susteren: I thought you were referring merely to the newspaper, or to the announcement of the department.

You will have to address that question to Mr. Robertson. He represents the Department of Agriculture. . . .

Mr. Yannacone: Does this report which you signed represent fairly and substantially the substance of your professional opinion on the subject matter contained therein?

A: Yes.

Mr. Yannacone: Fine. Now I will call upon the witness to produce that report as representative of the substance of his professional opinion.

Examiner Van Susteren: You will have to ask Mr. Robertson. He represents the United States Department of Agriculture, and Mr. Robertson will decide whether that material will be secured or presented, and not the witness. . . .

Mr. McConnell: If it please the Examiner, I would like to interrupt for a moment. I would like to direct the question to Mr. Robertson, and ask if he would furnish to this hearing a copy of the letter.

Mr. Robertson: And for the record, I will state that when I return to Washington, I will, number one, attempt to find this document; it is something over a couple of years old; I don't know whether it's available---

Mr. Yannacone: June 30, '67.

Examiner Van Susteren: Just a moment, you are interrupting, Counsel.

Mr. Robertson: If I may finish, Mr. Yannacone, without any remarks---

Mr. Yannacone: Excuse me.

Mr. Robertson: Number one, that would be the first determination. Then an evaluation will be made, and hopefully that document will be furnished to this hearing.

Mr. Yannacone: An evaluation by whom?

Mr. Robertson: To determine whether or not it is in compliance with departmental regulations to disclose the contents of this document. It is interoffice correspondence. There are regulations with respect to disclosure of that type of correspondence that I am not that well versed in, and I will admit it.

Mr. McConnell: Excuse me. Assuming there is no formulization material in this, might we assume that we would then be able to receive this document in due course upon your return?

Examiner Van Susteren: Well, first of all, the Examiner has ruled several times that the work of what this committee did and said and deliberated and so on, one, had no relevancy and materiality, and it merely represented [Dr. Hays's] professional judgment. We could be here for the next ten years probing Dr. Hays's mind.

Mr. Yannacone: Now, Dr. Hays---

Mr. Stafford: Is that a ruling?

Examiner Van Susteren: That's a ruling. If Mr. Robertson doesn't want to submit it, then it won't be submitted.

Mr. Yannacone: Dr. Hays, we don't want to probe your mind. I just want to ask you, will you summarize for the record now just what your professional scientific independent judgment is with respect to the subject matter of that investigation, [not] persistent pesticides in general, but only with respect to DDT in particular?

Mr. Stafford: Object to that on the grounds of relevancy.

Examiner Van Susteren: If the witness wants to render an opinion and so on, he may do so. If he does not want to give his professional opinion as requested, he need not do so.

Depends upon you, Dr. Hays, whatever you want to do.

Dr. Hays: I would rather not do so.

Examiner Van Susteren: All right.

Mr. Yannacone: You will not do so?

Examiner Van Susteren: Just a moment, Counsel. I warned you before I don't want you making remarks like this to the witness. This borders on badgering the witness. Now just ask your next question.

Mr. Yannacone: All right. Doctor, tell us why you don't want to render a professional opinion.

Mr. Stafford: Object to the question.

Examiner Van Susteren: And the objection is sustained. He has a right to have an opinion or not. And if he doesn't want to give an opinion---

Mr. Robertson: Mr. Examiner---

Examiner Van Susteren: ---then he isn't going to give his opinion.

Mr. Yannacone: Doctor, do you have an opinion?

Mr. Stafford: Object.

Mr. Yannacone: Maybe I should have asked you that first.

Mr. Stafford: Objected to for the same reason.

Examiner Van Susteren: If the witness has an opinion and he wants to state yes or no, he can answer it yes or no. However he desires.

Dr. Hays: Your Honor, you gave me a choice, and I chose the one that I gave you.

Examiner Van Susteren: So you are not going to express an opinion?

Dr. Hays: Since the Examiner gave me the choice of either responding or not responding, I chose not to respond.

Mr. Yannacone: Mr. Examiner, may I ask for a ruling from the Examiner that this witness should either render such an opinion or state he has no such an opinion. Since when is a witness who has testified on examination by Mr. Stafford as to [the] certain harmless[ness of] and protection of the populace from the hazards of DDT, permitted to say: I don't want to now render an opinion?

Examiner Van Susteren: His direct testimony concerned registration procedures. As to whether Dr. Hays has an opinion or not at the present time, is up to Dr. Hays. He stated that even if he had an opinion, he did not want to render it. And we will let it go at that.

Mr. Robertson: Furthermore, Mr. Examiner, Dr. Hays said he had signed this report, or whatever it was. It is a document that's two years old.

Mr. Yannacone: All right. Dr. Hays, has your independent professional scientific opinion, whatever that might be, changed since you signed that document?

Mr. Stafford: Mr. Examiner, this circuitous manner of evading the ruling---

Examiner Van Susteren: The objection is sustained.

Mr. Stafford: I object to any further questions along that line.

Examiner Van Susteren: The objection is sustained.

Mr. Yannacone: All right. Dr. Hays, is there any independent scientific evaluation of the data submitted by an applicant for registration or reregistration of a pesticide, in particular one such as DDT, made in the regular course of business of your department?

A: I think I so stated that the evaluation staff makes the review.

Q: Is the data, Doctor, on which the evaluation staff makes its review available for examination in the regular course of business of your department to outsiders who are not members of your department?

A: No, sir.

Q: Is it treated, to your knowledge, as privileged or confidential material?

A: This is treated as privileged and confidential.

Q: And it is not evaluated by any outside agency other than your technical staff of people at the levels you described?

A: Except the interagency people.

Q: Is the actual scientific data available to the interagency people?

A: Yes, sir. . . .

Q: Is it possible for any party other than the duly authorized representative of Health, Education and Welfare or the Secretary of the Interior, as set forth in this memorandum of understanding [previously mentioned in testimony], to review or examine those registration statements for DDT?

Mr. Robertson: Mr. Examiner, I believe we went into this this morning with respect to disclosure of information which is covered by certain federal laws and also departmental regulations.

Mr. Yannacone: Let's spell it out.

Mr. Robertson: Specifically is Mr. Yannacone asking the question: Can any outside person request this information?

Mr. Yannacone: There are two aspects of this memorandum of understanding, and I think we had better get it clear for the record. Dr. Hays, this understanding provides that each department will designate a scientist to act on behalf of such department in carrying out the terms of this agreement.

Who represents your department, Doctor?

A: Dr. Anderson.

Q: And who represents Health, Education and Welfare, if you know?

A: Dr. Kirk.

Q: And who represents the U.S. Department of the Interior?

A: Dr. Johnson.

Q: Now are the registration statements and the data submitted therewith available to anyone other than those named individuals, to your knowledge?

Mr. Robertson: Mr. Examiner, I believe he stated there is a free transmittal of information between the three agencies. Now is he trying to limit, in this question to Dr. Hays, that this information is only transmitted at this level of the three individuals that Dr. Hays has named? I would like to know specifically where we are, so Dr. Hays can answer the question.

Mr. Yannacone: I will explain it a little bit more clearly. . . .

Is it possible for anyone other than the named three individuals who are specifically authorized to act under this agreement in accordance with the provisions of Section 2.a. thereof, is it possible for any other individual in the U.S. Department of the Interior or the U.S. Department of Health, Education and Welfare to secure this registration data, the actual data on the registration of DDT and its related analogs? . . .

Is there anyone else? That's all I want to know.

Examiner Van Susteren: That he knows.

Do you know of anyone else who would have the authority?

Dr. Hays: No.

Examiner Van Susteren: All right, he doesn't know.

Mr. Yannacone: Okay. Now one more question.

Is it possible for any member of the general public to make an application to examine those documents? Is there any procedure in your department whereby such application can be made that you know of? . . .

Dr. Hays: Not to my knowledge, no.

Q: Is there any procedure whereby such information may be subpoenaed or made available; any request you have to honor other than the three-agency agreement here? . . .

Mr. Robertson: Mr. Yannacone, when letters come into the department or people come into the department requesting to examine certain documents on file with the department—and those documents necessarily are usually specified to some extent, because there are a number of documents—this matter is referred to a division within the Office of the General Counsel, a unit I am not in myself, to determine whether or not it would be in violation of the Public Information Statute, the Freedom of Information Law, and the Department of Agriculture regulations issued pursuant thereto. Once this determination is made on this particular request, the person so asking for the documents is informed, here are the documents, or is further informed that pursuant to the applicable provisions, the information cannot be furnished. But this request, if it came in to Dr. Hays or any other administrative official in the United States Department of Agriculture, would be referred through normal channels for this determination. . . .

Mr. McConnell: With regard to the cancellations that have taken place to date, have any reviews for cancellation purposes been instituted by your division, to your knowledge or since you have been there, for any purposes other than lethal effects upon the human population of this country?

Mr. Robertson: You understand the question, Doctor?

Dr. Hays: Yes.

   Not to my knowledge.

Q: And the two cancellations that did take place, were they both prior to 1966?

A: One was prior to '66. The other was in '68. . . .

Mr. McConnell: I have no further questions. . . .

Examiner Van Susteren: If there is nothing further of the witness, he will be excused.

   You are excused.

But Harry Hays wasn't really excused. Perhaps as a result of the bad showing he made at Madison, he was forced to make another public appearance before a congressional subcommittee in May headed by Congressman Fountain of North Carolina. At that hearing further data on the blatant inefficiency of the Pesticides Research Division came to light.

Whether the hearings in Madison and those in front of the congressional committee will lead to definitive changes in the procedures under which the public is protected from pesticides is questionable at this point, but what they certainly did was to open the eyes of many people to the machinations of the men who, through neglect, can poison the world.

11

The Conclusion

The evidence was in: competent witnesses had clearly outlined the disastrous physical properties of DDT, its persistence, solubility in lipids, broad biological activity, and surprising mobility. They had outlined the consequences of these properties: the fact that DDT was being stored in body fat and in the fatty layers of the nervous system; that DDT was not remaining restricted to the pests it was set out to eradicate but was also affecting beneficial insects, fish, and birds; and that concentrations of DDT could now be found throughout the biosphere from the phytoplankton of the oceans to the penguins of the Antarctic, and could be found in alarming proportions in human mother's milk.

Witnesses had testified about the physiological effects of DDT: They showed that DDT was inducing the liver to produce nonspecific enzymes which, in turn, could degrade steroid hormones or could interfere with the pharmacological activity of drugs administered as part of medical treatment. In addition, they showed that DDT was affecting the transmission characteristics of the nervous system and had been found to mimic the action of certain hormones.

Witnesses had directly linked biologically-concentrating DDT to the lack of reproductive success in fish, mortality in fish fry, population crashes in the golden eagle, Bermuda petrel, peregrine falcon, and other raptorial birds, and had linked the spraying of elms with DDT to nervous symptoms and population declines in robins.

With this evidence in hand, on May 21, 1969, almost six months after the first testimony had been given, Hearing Examiner Van Susteren adjourned the Madison proceedings and retired to prepare his findings of fact and conclusions of law, neither of which was published until a year later. The public did not deliberate as long; aroused by the overwhelming evidence so thoroughly broadcast by an awakened press, they demanded results. And results they got; in the Department of Agriculture;* in the Department of the Interior; in the

---

*The effect of the ruling by this department has been diluted by the subsequent appeal of the five largest manufacturers of DDT.

regulatory legislation of state, local and foreign governments; and in further law suits instigated by Yannacone and other concerned lawyers and environmental groups.

Therefore, many may feel that Van Susteren's ruling, published symbolically, on May 21, 1970, is anticlimactic. However, it does represent the first judicial consideration of the evidence for and against the continued use of DDT. It lacks the emotional involvement of the partisans and advocates at the courtroom drama, but substitutes the impartial retrospection and judicial review of the testimony of all witnesses. The ruling* follows:

DDT, including one or more of its metabolites in any concentration or in combination with other chemicals at any level, within any tolerances, or in any amounts, is harmful to humans and found to be of public health significance. No concentrations, levels, tolerances, or amounts can be established. Chemical properties and characteristics of DDT enable it to be stored or accumulated in the human body and in each trophic level of various food chains, particularly the aquatic, which provides food for human consumption. Its ingestion and dosage therefore cannot be controlled and consequently its storage is uncontrolled. Minute amounts of the chemical, while not producing observable clinical effects, do have biochemical, pharmacological, and neurophysiological effects of public health significance.

No acute or chronic levels of DDT which are harmful to animal or aquatic life can be established. For the reasons above set forth, a chronic level may become an acute level. Feeding tests, laboratory experiments, and environmental studies establish that DDT or one or more of its analogs is harmful to raptors and waterfowl by interfering with their reproductive process and in other birds by having a direct neurophysiological effect.

Feeding tests or experiments and environmental studies establish that DDT at chronic low levels is harmful to fish by reducing their resistance to stress.

DDT and its analogs are therefore environmental pollutants within the definitions of Sections 144.01 (11) and 144.30 (9), Wisconsin Statutes, by contaminating and rendering unclean and impure the air, land, and waters of the state and making the same injurious to public health and deleterious to fish, bird, and animal life.

With the federal government and the general populace alerted to the DDT problem and with Van Susteren's opinion registered, it would be easy to become cocky about the pesticide problem. But is this justified? Just what alternatives are there to DDT, and are these alternatives any "safer" than the chlorinated hydrocarbons?†

Many of the substitutes used today in place of DDT are organophosphates, compounds with the requisite short life span but with a drawback—they are highly toxic to mammals and to many lower

---

*The entire opinion appears in the appendix, pages 191 to 206.

†The following discussion is summarized from Robert van den Bosch, "Prescribing for the environment," *Pesticides,* A Scientists' Institute for Public Information Workbook (New York, 1970): 3–8.

animals including not only the insect pests they are designed to eliminate but also their predators. So, once again, the old problem returns, the pesticide is knocking out natural biological controls, is causing outbreaks of previously innocuous insects, and is hastening the development of resistant strains. The ecological problem here is clear; what may be less obvious and equally important to the farmer is the economic problem that arises as resurgence, eradication of beneficial insects, and secondary outbreaks force increasingly larger and more frequent doses of pesticides—at an increasingly higher cost.

Obviously, what is needed is the development of pesticides with greater ecological selectivity. However, this will be difficult as such schemes directly oppose the present aims of the pesticide industry: to produce a product with as wide use as possible so that development and marketing costs can be met and a reasonable profit be made.

Better recommendations for pesticide use form another element in the environmental prescription. At present, pesticide industry salesmen, by necessity, serve as "diagnostician, therapist, and pill dispenser" often without the technical competence necessary to perform any one of the functions adequately and always with an economic conflict of interest that would make dispassionate analysis impossible.

Finally, more enlightened procedures for screening and registering pesticides are needed to replace the present narrow standards of performance (killing efficiency) and safety to human health (the length of time the pesticide remains on a particular crop and the hazard that its residues will cause). In broadest terms, registration officials must know whether the percentages of insects killed are economically justified, and whether the problems engendered by the use of the pesticide will be greater than those solved.

Thus, it is obvious that new legislation is necessary if we are to accommodate the needs of modern society for intensive agricultural practices with high yields at low cost, and at the same time protect our fragile biosphere. Obviously, chemical pesticides have a place in modern agriculture but because they are inherently toxic they must be regulated. Therefore the challenge facing modern government is to develop ecologically sophisticated, socially relevant, politically feasible environmental legislation.

APPENDICES

APPENDICES

The Ruling

DEPARTMENT OF NATURAL RESOURCES
STATE OF WISCONSIN

Petition of
CITIZENS NATURAL RESOURCES ASSOCIATION, INC.,
WISCONSIN DIVISION, IZAAK WALTON LEAGUE
OF AMERICA, INC.

For a declaratory ruling on the use of dichloro-diphenyl-trichloro-
ethane, commonly known as DDT, in the state of Wisconsin

Docket 3-DR

EXAMINER'S SUMMARY OF EVIDENCE AND PROPOSED RULING

On October 28, 1968, the Citizens Natural Resources Association,
Inc. by Frederick L. Ott, Wauwatosa, Wisconsin, and on November
1, 1968, the Izaak Walton League of America, Inc., Wisconsin Division,
by J. Michael Borden, Elm Grove, Wisconsin, filed petitions with the
Department of Natural Resources requesting a declaratory ruling in
respect to the use of Dichloro-Diphenyl-Trichloro-Ethane, commonly
known as DDT, in the State of Wisconsin.

The Department issued its Notice of Hearing on November 5, 1968
and held hearing on the matter December 2, 1968 and on days thereafter at Madison, Wisconsin, before Examiner Maurice H. Van Susteren.*

---

*The list of Appearances which precedes the Examiner's Summary of Evidence has
been deleted. (Eds.)

Examiner's Summary of Evidence

I. Statutes, Rules, Issues.

A. The petitioners seek a declaratory ruling under Section 227.06, Wisconsin Statutes, which provides that any interested person may petition for a declaratory ruling with respect to the applicability to any person, property or state of facts of any rule or statute enforced by it. A ruling is sought declaring DDT to be a highly toxic persistent chemical, that its use be restricted in such way that it cannot enter the biosphere and that its existence in the biosphere constitutes pollution.

Section 144.01 (11), Wis. Stats., defines pollution:
Pollution includes contaminating or rendering unclean or impure the waters of the state, or making the same injurious to public health, harmful for commercial or recreational use, or deleterious to fish, bird, animal or plant life.

Section 144.30 (9), Wis. Stats., defines environmental pollution:
Environmental pollution means the contaminating or rendering unclean or impure the air, land or waters of the state, or making the same injurious to public health, harmful for commercial or recreational use, or deleterious to fish, bird, animal or plant life.

B. Rules
Under the provisions of Section 144.025 (2) (b), the Department is authorized to adopt rules setting standards of water quality, to protect the public interest which includes the protection of the public health and welfare and the use of the waters for public and private water supplies, the propagation of fish and aquatic life and wildlife and other uses.

In compliance with the above, the Department adopted minimum standards of water quality in Wisconsin Administrative Code section RD 2.02 (1) (d):
Substances in concentrations or combinations which are toxic or harmful to humans shall not be present in amounts found to be of public health significance, nor shall substances be present in amounts, which by bio-assay and other appropriate tests, indicate acute or chronic levels harmful to animal, plant or aquatic life.

C. Issues.
1. DDT in what concentrations or combinations is toxic or harmful to humans and its presence in water in what amounts can be found to be of public health significance?
2. What amounts of DDT in water, which by bio-assay and other appropriate tests, indicates acute or chronic levels harmful to animals, plant or aquatic life?
3. Is DDT a pollutant within the statutory definitions of "pollution" as found in Sections 144.01 (11) and 144.30 (9) of the Wisconsin Statutes?

II. DDT—Chemical Structure

A. Chemical Structure

DDT, a chlorinated hydrocarbon, is chemically described as 1, 1, 1-trichloro-2, 2-*bis* (para-chlorophenyl) ethane.* It is common practice to refer to DDT and to include in the term its various isomers—DDE and DDD, and lesser isomers DDA and DDMU. Each of the isomers is further broken down into the ortho, para derivative and the para, para derivative. Isomers have the same number and kind of atoms but in either different configuration or different locations within the molecule. What is known as technical DDT is composed of 75% of para, para DDT, 20% ortho, para prime DDT, and 5% is the other isomers. DDT is manufactured in the reaction of chloral and monochlorobenzene in the presence of sulfuric acid, the spent sulfuric acid carrying away any monochlorobenzene that remains.

DDT has been known since 1870, developed in Switzerland as an insecticide and first used in United States armed forces in 1943. DDT in its pure state is not insecticidal.

B. Properties, Uses, and Tests

1. Physical Properties

DDT is unique in that it has broad biological activity as an insecticide, it has great chemical stability, high mobility, is relatively insoluble in water and soluble in fat or lipid tissue. It has a strong tendency to form suspensions more than solutions. Ability to absorb particulate matter can cause a concentration in water particulates which is thousands of times greater than the concentration in water itself. Although it has a finite vapor pressure, and is relatively non-volatile, it will evaporate and recrystallize in air and travel in airborne dust. It also has the power of co-distillation. Half-life is estimated at ten years, there being no standards of determination.

However stable DDT is, it does degrade in sunshine and oxygen, first to DDE and then to bis (chlorophenyl) ketone and parachlorobenzoic acid. Since the DDT molecule is a very efficient light absorber, it absorbs the shorter wave lengths of the solar spectrum. Being a diphenyl methane type compound, it readily dissociates to a free radical leading to the formation of DDE. The degrading mechanism of DDE is not clear, but it does degrade faster than DDT. Under sunlight irradiation in sealed Petrie dishes, DDT will degrade approximately 50% in 12 days and it can be expected to be faster in a vapor state in the atmosphere.

When applied to crops, it has a period of rapid decline believed to be evaporation, and then degrades at a rate that is logarithmic

---

*References to specific pages in the hearing transcript have been deleted throughout this document. (Eds.)

in time. Exact mechanism of "disappearance" is not known. Half-
lives of DDT and isomers are reported in crop materials from
as little as two days to 40 days—the longer time being the more
appropriate. DDT on crops or plants degrades to DDD and DDE
and when harvested, the residues go with the plant. DDT very
tightly sorbs to soil particles which may be moved deeper into
soil or washed away. "Disappearance" of DDT from soil is logarithmic in time and on the average 20% of the amount present
at any time "disappears" each year. DDT "degradation" in soil
is due to microorganisms amounting to as much as two-thirds
of the DDT in a period of two weeks to DDE, DDD, DDA, DBM,
DBP and chlorobenzoic acid. Degradation is more rapid under
anaerobic conditions. DDT is extremely insoluble in water, sorbs
to particulate matter which collects on bottoms of streams, etc.
and tends to degrade under the anaerobic conditions existing.
It will degrade in the gut of fish, insects, mammals and birds
because of microorganisms found there.

2. Uses

DDT is and has been primarily used as an agricultural pesticide,
for the control of Dutch elm disease and for the control of
mosquitoes. Slightly over 100,000 pounds were shipped into
Wisconsin in 1968 with approximately 58,000 pounds used for
Dutch elm disease. Recommendation for mosquito control was
withdrawn (in Wisconsin), as was registration for use on dairy
cattle and around dairy barns since a zero tolerance was established for milk in 1954 in the Miller amendment. It is also used
as a rodenticide in the control of mice and bats.

3. Toxicity

DDT has a wide range toxic effect on agricultural pests,
on the entire phylum Arthropoda, and is toxic to a degree
to fish, birds, mammals, amphibians and reptiles and crustaceans.
It is a Class III pesticide under the Federal Economic Poisons Law.
Technical DDT is somewhat more toxic than metabolites DDE
and DDD and toxicity varies between species. Pure undiluted
DDT is not an insecticide. It was stipulated that DDT has an effect
on the nervous system. Effects occurring at very low concentrations are variable but in the case of cockroaches consist of
repetitive firing in nerves, resulting in disorientation, running
about, tremor, overwork, and death due to exhaustion.

The effect on crustacean nerve is the failure of the nerve axon
to transmit impulses. The effect on robins is similar to that of
the cockroach—quivering, tremor, and death due to nerve effect.
DDT, at acute toxic levels, affects the central nervous system of
fish causing instability, difficulty in respiration, and sluggishness.
In humans, the first clinical signs of toxicity are a sensation of
burning or itching of the tongue, lips, and face, tingling of fingers
and toes. With larger doses tremor appears, a sensation of alarm,

of fear, marked uneasiness, and convulsions with accidental doses.

In rats clinical signs of poisoning are directly correlated with the concentration of DDT in the nervous system, as measured by the concentration in the brain. All parts of the nervous system are affected. In both animals and man poisoned by DDT it is the nervous system that is affected primarily.

4. Translocation

DDT, because of its chemical/physical properties, once applied to crops or placed in the atmosphere, moves throughout the environment in water, air, soil, and food. A minor transport mechanism is in organisms such as birds and fish. It has been found in filtered air, untreated forest soil, and the fish in untreated watersheds had in some instances 2.4 ppm DDE. DDT and metabolites have been found in oceanic food chains from zooplankton to gulls, osprey, cormorants, petrels, pelicans and peregrine falcons with a corresponding biological concentration at each level of the chain. It was not present in plankton gathered in the 1920's and studied, nor in the body of a penguin of 1911. It was not present in ten samples of human fat collected before the advent of DDT. DDT is found in bodies of fresh water (Lakes Michigan/Superior) from the muds up through all aquatic organisms.

5. Storage/Accumulation in Organisms

While relatively insoluble in water, DDT is soluble in lipid or fatty tissue, and accumulates in such tissue.

Unlike birds, DDE levels in mammals appear to reach a plateau with the same amount of intake and remain at approximately 10-12 ppm in the body fat of the general population. A study of 35 men with 11 to 19 years of exposure in a DDT manufacturing plant shows the overall range of storage of isomers and metabolites in body fat ranged from 38 to 647 ppm, with an average of eight ppm for the general population. There is no evidence of progression of storage of DDT in the general population since 1950–1951.

Abstainers from meat stored on the average about half as much DDT and half as much DDE as persons in the general population who stored an average of 4.9 ppm and 6.1 ppm of the two compounds respectively. Agricultural occupational exposure caused marked increase in storage at an average concentration of 17.1 ppm. In a 1958 study applicators stored less than formulators but stored about three times as much DDT/DDE as neighbors with only environmental exposure. A 1961–1962 study showed no difference in levels of DDT/DDE between the general population and persons living in or near areas of extensive agricultural use. Differences in the two studies may be due to decreased use by more than 50% in the test area. One of the general population samples came from a six-month-old bottle-fed boy whose extractible lipid levels for DDT (6.2 ppm) and DDE (16.5 ppm) were well within range of the general population. In six months he had stored at least the average concentration for the general population.

Human storage of DDT in England, West Germany, France, and Canada is lower than in the United States. Only in Hungary did the level of DDT storage equal or exceed that of the U.S. Sixty-five per cent of the material stored is DDE. Loss of DDT from storage, although always slow—less than 0.3% per day in man—is always more rapid when the storage is high than when it is low. Rate of excretion of DDT changes with concentration. In general, doubling the dosage will double the storage. A total of 282 autopsy samples of human abdominal fat tissue obtained at random from patients who had died of a variety of causes, regularly showed concentrations of 0.1 ppm or higher. The sum of DDE plus DDT averaged 10.3 ppm with standard deviation of 7.2 ppm. Nearly 96% of the values ranged from 0.1 to 22.3 ppm. About 72% of the DDT was present as DDE, the rest as DDT itself. Other studies of volunteers receiving 35 mg per man per day of technical DDT resulted in average storage in fat of 234 ppm and 281 ppm respectively. Other studies of 1961 to date give ranges of storage at between 2.3 and 4.0 ppm body fat, but the witness was uncertain. Concentration or storage depends on the dosage and the nature of the tissue in which stored and the rate of metabolism depends on the concentration. A point or plateau is reached where concentration/storage and rate of metabolism matches ingestion or dosage.

All organisms store DDT in varying amounts in fat, muscle, and internal organs.

6. Detection and Measurement

The laboratory equipment now most commonly used for the detection and measurement of chlorinated hydrocarbons is what is known as the gas chromatograph. Long U-shaped tubes filled with a powder-like material coated with resin are in a heat cabinet with temperatures around 200° C. A specific amount of a mixture of various chlorinated hydrocarbons in a solvent such as hexane is injected through the system at one end of the column. Some of the compounds come out earlier than others because of their physical-chemical properties and pass through a mechanism known as a detector, the most commonly used being known as an electron capture detector. The detector operates a recorder pen which marks a moving sheet of paper showing the times and amounts of gas which spin off the basic hydro-carbon ring. Peaks are made on the moving paper corresponding to the compound coming off. Because of chemical properties it occasionally happens that two of the compounds may come out at or about the same time, creating one peak or confusing the type of peak. This may happen with a nonpolar column known as a DC-200 or a SE30 where compounds known as polychlorinated-biphenyls may come off at the same time as *p,p'* DDT. The interference is avoided by also using a QF-1 polar column where the retention times of the compounds will be differentiated. Laboratory procedures such as saponification can also be utilized. The chlorinated hydrocarbons are extracted from the sample provided, no matter what it may be, through certain laboratory procedures.

Almost all analytical techniques are based on pesticide manuals of the Food and Drug Administration listing techniques and how to verify conclusions. There are several techniques and procedures used for confirmation or verification.

Chlorinated hydrocarbon residues in organisms or samples of other material can be detected and measured in parts per million (ppm) up to .10 ppm in the Schecter-Haller colorimetric method and with the more sensitive gas chromatographs up to parts per billion (ppb).

7. Chlorinated hydrocarbons are tested for effects, storage, and residues by feeding tests, bio-assay, in "vitro" and in "vivo."

III. Enzyme Induction

A. Enzymes—Function and Body System Effects

1. Enzymes and Liver Function

Enzymes are catalysts, protein in nature, manifested morphologically by an increase in the smooth *endoplasmic reticulum* of the liver. The production of nonspecific hydroxylating enzymes by the *microsomal* fragment liver cells is a normal function of the liver. The existence of nonspecific enzymes and their production in the liver became evident in 1958. Two major functions of the liver are digestion and detoxification. Enzymes are produced in the liver to *hydrolyze* or *oxidize* non-polar substances such as fat and resins to a polar condition, capable of excretion by the kidney. They have not been isolated in purified form. They will also hydroxylate both endogenous and exogenous steroids which is a natural physiological process and will themselves disappear when no longer needed. The enzyme substrate affinity for steroids is much greater than it is for drugs and it has been shown that *o,p'* DDD stimulates the metabolism of hydrocortisone to 6-Beta hydroxy-cortisone in humans. Enhanced steroid metabolism causes compensatory synthesis of more steroid by organisms and raises problems of drug interaction.

Many drugs, compounds, and poisons induce detoxifying enzymes. For an example, while phenobarbital induces enzymes and is metabolized, a high level of enzymes induced by chlorinated hydrocarbons will decrease its physiological effect. Levels of enzymes induced are directly proportional to the dose and rate.

An overall body concentration of one ppm and ten ppm of DDT in the fat—levels slightly below levels found in the human body—causes a significant increase in liver enzymes. Five ppm of DDT given to rats for three months caused an increase in enzyme activity in the rat liver and as little as 40 micrograms of DDT given to rats for four weeks caused an increase in the metabolism of several clinically useful drugs, an effect which extended for twelve weeks after the DDT feeding was stopped. Liver detoxification mechanisms are less highly developed in human infants than in adults and this is also true of rat infants.

2. Clinical Effects

There are no overt pathological changes in body organs at tissue levels of ten ppm. There are no overt pathological *histological* changes in rats fed DDT in very low amounts. This does not mean, however, that there are no bio-chemical or pharmacologic effects. Histologically, no effect is evident except a proliferation of the smooth endoplasmic reticulum of the liver cell following the administration of enzyme-inducing agents. Up to this time this effect has not been considered a pathological one and therefore has no pathological significance. The biochemical significance is a higher enzyme level and the metabolism at a faster rate of steroids and drugs.

Repeated oral doses of DDT to volunteers of approximately 200 times what the general population gets in their diet caused no clinical signs of illness. Dosages of 3.5 and 35 mg, depending on the exact weight of each man, averaged .05 and .5 mg/kg per day and did not produce any clinical illness. It did result, however, in increased storage. Storage in fat of 35 mg per man per day dosages of technical DDT was an average of 234 ppm in one study and 281 in another study. Average person's diet intake of DDT is .028 mg per man per day on values established for 1964 through 1967. Ten mg/kg produces illness in many people and sixteen mg/kg has frequently produced convulsions. Routine clinical examinations and simple laboratory tests of workers with great exposure to DDT failed to show any illness and disorders were looked for which had been produced by moderately larger dosages of DDT in man and in animals. There were no clinical or sub-clinical effects found. Clinical effects were primarily sought. No procedures were utilized to determine sub-clinical effects of enzyme induction. Not only the dosage that ordinary people get but the higher dosages workers get produce no detectable clinical effects. People who have not ingested enormous amounts of DDT have evidenced no manifest clinical symptoms.

3. Sub-clinical Effects

The principal sub-clinical metabolic effect of DDT is the accelerated degradation of drugs, hormones, and *endocrines* as listed in preceding paragraphs. Hormones such as testosterone, estrogen, and progesterone in the human body function in a ratio of parts per billion (ppb) and are broken down by liver enzymes. Other chlorinated hydrocarbons such as polychlorinated biphenyls (PCB) also induce hepatic enzymes with the capacity to hydroxylate steroids. The PCB-induced enzyme produces a different polar metabolite. PCB is found in almost all organisms throughout the world.

The degradation of hormones without observable effect on the organism itself by DDT-induced hepatic enzymes is evidenced by what is known as "the thin eggshell phenomenon" in birds. High levels of hepatic enzymes cause a rapid metabolism of the avian estrogen necessary for the laying down of calcium in the medullary bone of birds. An upset in the calcium metabolism of the bird itself results in abnormal brooding behavior and the eating of its own eggs. Lack of available calcium carbonate results in thin eggshells.

In a controlled feeding study at the Patuxent Wildlife Research Center, Department of the Interior, mallards were fed DDT or metabolites DDD or DDE each separately at 10 ppm or 40 ppm on a dry weight basis. Mallards fed DDE cracked or broke 24% of their eggs, controls 4%, and good eggs produced had shells 13½% thinner than controls. Incubated eggs laid by ducks fed DDE produced less than one-half (½) as many healthy ducklings as did the controls. Both amounts of DDE produced effects of similar magnitude. A "no effect level" was not found and must be lower than the dosages given. Ducks fed p,p' DDT minus the o,p fraction at 25 ppm showed results similar to those produced by DDE at 10 or 40 ppm. DDD fed to ducks at 10 or 40 ppm produced normal eggs and healthy ducklings.

In another experiment, kestrels (sparrow hawks) were fed 2 ppm of DDT plus ⅓ ppm of dieldrin and also 5 ppm DDT plus 1 ppm of dieldrin. Shells of dosed birds were 15% thinner than the controls as were second generations fed diet of the parents.

Another experiment with the Japanese quail, was designed to determine what effect toxic levels of DDT fed in the growing period might have on reproduction efficiency later on, and whether resistance to DDT could be developed. The quail were fed 200 ppm until a 50% mortality was reached which was approximately 30 days. Survivors were bred for four generations. There was no effect on reproduction or shell thickness. No residue levels were determined. The Japanese quail is related physiologically to chickens, pheasants, and turkeys.

What role, if any, DDT plays in fish physiological processes is unknown. It is unknown whether the principle of hepatic enzyme induction is involved in fish reproduction.

B. Environmental Effects
1. Birds
   a. Sea or Waterfowl
   The effects shown in the Patuxent studies are similar to those experienced in the environment, resulting in declining raptor and water bird populations. The pelagic Bermuda petrel, the brown pelican, the peregrine falcon, the forked-tail petrel, the bald eagle, the osprey, the coopers hawk, the double-crested cormorant, and mallards are affected.
   b. Upland Birds
   The population trend on a nationwide basis so far as small game and upland game birds are concerned, taking into account anticipated population fluctuations, shows a generally healthy upward trend. This is true of cottontails, snowshoe hare and squirrels and birds such as pheasant, quail, grouse, and woodcock.
   c. Omnivores
   Golden eagles, red-tailed hawks, and the great horned owl are omnivores, feeding on both birds and mammals, and representative of reasonably stationary populations showing no statistically significant fluctuations in eggshell weight. The golden eagle in the British Isles is also declining and it is shown that in feeding on sheep carrion and fleece is ingesting DDT in combination with dieldrin, both being used as sheep-dips.
2. Fish
   The University of New Hampshire studies on effects of DDT on reproduction of brook trout and their resistance to stress show that mortality where one gamete came from fish exposed to DDT was in every case higher than instances where neither of the gametes came from a fish exposed to DDT, and in five out of six cases the differences were statistically significant. The stress imposed was starvation and lowered water temperatures and the approach of spawning activity.
   The findings in the New Hampshire study are consistent with the study and field investigation of the New York State Conservation Department, which showed a relatively close relationship between the amount of DDT in lake trout eggs and observed mortality. Canadian studies establish that when DDT and its metabolites occur in concentrations slightly above 400 ppb in eggs, mortality in the resulting fry ranged from 30% to 90% in the 60-day period following the swim-up stage. The Wisconsin studies, however, are inconclusive. Data indicated no relationship between pesticide residue levels in eggs and median life span of fry nor pesticide levels in fry and median life span of the fry. Difficulties in the experiment occurred and the experiment design differed also from the New York and New Hampshire studies.

IV. Effect on Nerves

  A. Anatomy and Physiology

Exhibit 184 is a simplified diagram of a motor nerve cell showing receptor *dendrites*, the nerve soma or body, the axon and endplate. The axon or transmission part of the nerve cell is composed of a central core surrounded by a plasma membrane called the neurolemma. Next to the neurolemma is a Schwann cell which comprises the myelin sheath. The myelin sheath has openings or channels called the node of Ranvier.

The neurolemma is believed to be a laminated membrane with a lipid molecule interior sandwiched between layers of protein molecules with channels through the membrane. The term "channel" does not refer to an anatomical structure, but only refers to a conceptual pathway. Between the neurolemma and the cytoplasm is a concentration of potassium *ions* and on the outside of the membrane is a concentration of sodium ions. The membrane offers resistance to the passage of electrical currents and shows selective permeability to the exchange of the sodium and potassium ions. In the resting cell the inside of the membrane is negative to the outside.

In excitation of the cell, the nerve impulse received by the dendrites at a certain potential is passed on through the axon. With the decay of the potential due to resistance, sodium ions pass through the membrane to the inside and simultaneously the potassium ions flow outward. Both flows last but a thousandth of a second and at the moment the two effects balance one another, the inside of the membrane is slightly positive and the membrane restores ionic balance typical of the resting state. The excitation or nerve impulse passes down the lattice-type membrane by leaking currents through one node of Ranvier to the next node where sodium and potassium ions are exchanged through the neurolemma. The conductance is then not a smooth flow but a whole series of changes in potentials. (Examiner's Note: This is a gross oversimplification of an enormously complex, complicated neurophysiological reaction.)

DDT will increase membrane conductance to potassium or the inactivation of the nerve membrane conductance to sodium or both are inhibited, thereby increasing the negative after-potential and its prolongation. DDT does not simply reduce maximum sodium current, but instead gums open the channels or some of the channels for sodium, resulting in the prolonged action potential and the prolonged action current with an effect similar to *veratrine*, the alkaloid active ingredient of belladonna.

> . . . A prolongation of the active state of the nerve means—and coupled with no change in the inactivation process for sodium— that after one nerve impulse, when the channels become ready to conduct or to open again to conduct a second impulse, and examine or test the potential across the membrane, they will find

that according to the potential they should already go again, that is, no further impulse would be necessary to make it fire another signal. This could produce repetitive firing. On the other hand, in other nerves where sodium inactivation takes a longer time, one could simply find that the nerve on reactivating its sodium mechanism finds that the potential is already too high to fire an action potential, and instead would simply remain quiescent. This would produce a complete failure of the transmission line or at least intermittent failure. Either of these mechanisms quite clearly and very conclusively could cause tremors and could cause grave disturbances in terms of the ability of the animal to move or to make motions.

The effects of DDT on nerves involved in the experiments conducted did not wear off during the time course of the experiment. The effects are irreversible. There did not seem to be any concentration lower limit of DDT to create the effects shown. In cockroaches, insects, and crustaceans, effects occur at very low concentrations and are variable. Effect on the cockroach consisted of repetitive firing in nerves where one impulse applied to the nerve no longer evoked a single message but rather a large volley of messages. The behavior showed disorientation, running about, tremoring, kicking of legs in the air, and death. There is a similar effect on robins. How DDT reaches the nerves in the body is unknown.

It was stipulated that DDT has an effect on the nervous system. It is the nervous system that is primarily affected, and it does in fact stimulate all parts of the nervous system. The earliest subjective clinical sign of DDT poisoning is tingling of fingers and toes, and peculiar tingling about the mouth.

B. Environmental Effects

1. Fish

At acutely toxic levels, the chlorinated hydrocarbons damage the central nervous system, causing instability, difficulty in respiration, and sluggishness in fish.

2. Birds

In Hanover, New Hampshire, a study was made of effects of DDT on local birds after the town was sprayed for Dutch elm disease. Ninety-six dead birds had a concentration of about 30 ppm DDT. All birds with tremor had above 30 ppm in the whole bird.

The study conducted at the East Lansing Campus at Michigan State University shows that birds, primarily robins, suffered from extensive tremor and death after the campus had been sprayed with DDT. More than 200 robins and 216 specimens of non-robins, representing about 50 different species, were analyzed. It was determined that the amount of DDT in the liver had no correlation with mortality. The brain was considered the best criterion. DDT was found in 99.5% of birds from DDT-sprayed area. No robins with tremor were found that did not have 50 ppm or more DDT. All birds in tremor and dying birds had large concentrations of DDT in the brain.

3. Mammals

The giant squid *axon* fiber is like nerves of higher vertebrates, but unlike them has a very large diameter and is useful in nerve conductance experiments. Work done on the squid axon is applicable to nerves of both vertebrates and invertebrates. The effect of DDT on nerve conductance would not always be manifest in gross neuromuscular clinical signs and if manifested would not necessarily be the sole cause of the observed clinical signs.

The action of DDT in animals is manifested almost entirely through the nervous system with prominent signs of poisoning being muscle tremor, uncoordination, and convulsions. Concentration of DDT in the brain of rats fed *p,p'* DDT at 200 ppm for 90 days increased during a subsequent 10-day period of partial starvation. Increased concentration of DDT in the brain during starvation was correlated with clinical signs of poisoning. While all parts of the nervous system are affected by DDT the brain is of major importance.

V. Other Effects

A. Hormone Mimicry in Quail and Rats

An experiment designed to determine estrogenic effects resulted in a tripling of the wet weight of the coturnix quail oviduct when injected with 190 mg/kg of *o,p* DDT. No attempt was made to determine the amount that would cause the effect. The coturnix is a gallinaceous bird, related physiologically to chickens, pheasants and turkeys. Experiments at the University of Wisconsin designed to determine the effect toxic levels of DDT fed in the growing period would have on reproductive efficiency later on, and also whether DDT-resistant birds could be developed, showed no effect at all on coturnix eggshell thickness. The nationwide trend so far as small game and upland game birds are concerned shows "a generally healthy upward trend."

DDT also increases the uterine wet weight in immature female rats and in ovariectomized adult rats. The physiological significance of this effect is unknown.

B. DDT and Human Pathology

DDT concentrations were determined at autopsy in the fat and liver of 271 patients previously exhibiting pathological states of the liver, brain, and other tissues and compared with other random autopsy cases. There was a striking lack of correlation between concentrations in the liver and fat in all cases, but a significant correlation between levels in the brain and fat. There was no elevation of concentrations in the presence of brain tumors but a significant increase of the mean *p,p'* DDE in encephalomalacia and cerebral hemorrhage. Significant concentrations of *p,p'* DDE were found in portal "cirrhosis" and highly significant concentrations in carcinoma. Fat concentrations of the various DDT analogs were consistently and significantly elevated in hypertension.

Individuals using pesticides extensively in the home had levels of *p,p'* DDT and *p,p'* DDE three to four times higher than those who had used minimal quantities. No conclusions can be drawn on the role of pesticide exposure in the production of the diseases without confirmatory studies to determine whether the diseases caused high DDT levels or vice versa.

Opinion

Clinically observable toxic effects of DDT in humans are obtained only with extremely large dosages by sudden extreme exposure, or of accidental origin. Clinically observable effects are evident injury, illness, loss of body function which directly inconveniences a person at work or play. Toxicity, as the word is ordinarily expressed, is related to dosage which in turn is related to storage.

DDT is ubiquitous. It is found in the atmosphere, soil, water, and in food in what might be considered minute amounts. The chemical property of being soluble in lipid or fat tissue results in storage primarily in the body fat and nervous systems of all organisms in all levels of food chains. It is therefore impossible to establish levels, tolerances or concentrations at which DDT is toxic or harmful to human, animal, or aquatic life.

The principle that DDT, being a chlorinated hydrocarbon, induces the production of non-specific detoxifying hydroxylating hepatic enzymes is well established. The induction of the enzymes is a normal adaptive hepatic process for the detoxification of substances and no definitive pathological effects are observed at present dosages. A high level of induced hepatic hydroxylating enzymes, however, causes a pharmacological biochemical effect in accelerating the metabolism of body steroids and drugs such as barbiturates and nonbarbiturate depressants.

While the exact physiological mechanisms are not known in enzyme induction, it is established by feeding studies that DDT and one or more of its metabolites will by themselves cause a thinning of egg-shells in raptor, pelagic, or waterfowl birds. The effect explains the existence of the phenomena in the environment but does not exclude other causative factors, namely diet, illness, and other chlorinated hydrocarbons among them being polychlorinated biphenyls. The appearance of the phenomena, however, on two continents simultaneously, would seem to eliminate illness, diet, or predator interference as causative factors. Waterfowl and raptors on the top of water and other food chains are suffering decline and insect/worm-eaters are affected whereas gallinaceous birds are not. The differences in dosage reactions can also be explained by well-known order differences in birds.

The effect of DDT in minute amounts on the extremely complex, complicated mammalian nerve system is unknown. Huge dosages of DDT bordering on the accidental will cause gross clinically observable neurological symptoms in humans. It is uncontroverted that DDT has an almost immediate nerve effect on the primitive nerve systems of insects and on the less well-developed nervous systems of other forms of life. It is also uncontroverted that nerve tissue of vertebrates and invertebrates is the same, that DDT has a harmful effect on nerve conductance as shown by experiments on the axon of crustaceans and amphibians, that the effects are irreversible during the duration of the experiments. Clinically observable signs of nerve effects in humans such as tremor disappear upon reduction of dosage. That there are sub-clinical residual effects can only be postulated on mathematical equations and principles worked out in conjunction with nerve conductance experiments on nerve axons of crustaceans and amphibians and shown to be valid in all cases. Taking into consideration the above experiments together with the fact that DDT is used as a rodenticide for mice and bats, the only valid permissible inference is that DDT in small dosages has a harmful residual effect on the mammalian nervous system.

While the physiological mechanism causing a reduced reproductive success in fish and a reduced resistance to stress when dosed with fairly high levels of DDT is unknown, the known effects themselves can only be considered harmful ones.

The record is replete with evidence of the economic benefits derived from use of DDT in the control of pests in agriculture and in the control of mosquitoes for both comfort and prevention of disease. Without doubt DDT has provided enormous economic benefits, but economic benefits are not an issue or part of any issue in this case.

Ruling

DDT, including one or more of its metabolites in any concentration or in combination with other chemicals at any level, within any tolerances, or in any amounts, is harmful to humans and found to be of public health significance. No concentrations, levels, tolerances, or amounts can be established. Chemical properties and characteristics of DDT enable it to be stored or accumulated in the human body and in each trophic level of various food chains, particularly the aquatic, which provides food for human consumption. Its ingestion and dosage therefore cannot be controlled and consequently its storage is uncontrolled. Minute amounts of the chemical, while not producing observable clinical effects, do have biochemical, pharmacological, and neurophysiological effects of public health significance.

No acute or chronic levels of DDT which are harmful to animal or aquatic life can be established. For the reasons above set forth, a chronic level may become an acute level. Feeding tests, laboratory experiments, and environmental studies establish that DDT or one or more of its analogs is harmful to raptors and waterfowl by interfering with their reproductive process and in other birds by having a direct neurophysiological effect.

Feeding tests or experiments and environmental studies establish that DDT at chronic low levels is harmful to fish by reducing their resistance to stress.

DDT and its analogs are therefore environmental pollutants within the definitions of Sections 144.01 (11) and 144.30 (9), Wisconsin Statutes, by contaminating and rendering unclean and impure the air, land, and waters of the state and making the same injurious to public health and deleterious to fish, bird, and animal life.

Dated at Madison, Wisconsin, this __21st__ day of __May__, 1970.

STATE OF WISCONSIN

Model Pesticide Law

SECTION 1. Preamble

Since pesticides are useful in the control of certain insects, weeds, fungi and other forms of plant and animal life which have caused significant damage to man and his interests, but at the same time pesticides may contaminate the environment and have undesirable ecological effects, and therefore should be used to augment natural controls, it is deemed necessary for greater protection of the public health and welfare and to insure environmental quality consistent with the benefits derived from the safe and proper application of pesticides, to establish a pesticide control board that formulates pesticide policy in the ......(municipality)...... and administers and coordinates state efforts to control the use of pesticides in the (municipality)

SECTION 2. Definitions

In this act:
    (1) *"Board"* means the pesticide control board.
    (2) *"Control"* means maintaining pest population density at or below the economic threshold.
    (3) *"Economic threshold"* means the pest population density above which there is significant damage to man or his interests.
    (4) *"Non-target organism"* means any organism which the particular pesticide is not intended to control in a given application.
    (5) *"Pest"* means any organism that is present at a population density above the economic threshold.
    (6) *"Pesticide"* means any substance or mixture of substances intended to control pests and includes those substances commonly referred to as insecticides, fungicides, herbicides, and rodenticides.
    (7) *"Target organism"* means any organism which the particular pesticide is intended to control in a given application.

SECTION 3. Pesticide Control Board

(1) Appointment, membership. There is hereby created a pesticide control board consisting of 7 members, to be appointed by the ....(municipal executive)........ with the approval of the (municipal legislature) .... The appointments shall be made in writing and filed in the office of the ....(municipal executive)........ . The term of each member shall be the three years next following January 1 of the year in which his appointment is made and until the appointment of his successor except that the first six members shall be appointed respectively for such terms that on January 1 in each of the three years next following the year in which they are appointed the terms of two members will expire and except that a member appointed to replace a member who did not complete his full term shall be appointed for the balance of the term. Insofar as possible the board shall include appointed members representative of each of the following fields: aquatic or marine biology, agriculture, ecology, entomology, fish or game management, and pesticide application.

The following or their delegates shall serve as *ex officio* members of the Board: ....

....

....

(2) Organization

(a) The municipal executive shall call the first meeting of the board at the municipal hall without delay. The board shall elect from its membership a chairman and a secretary who shall serve for two years terms ending on January 1 next following a municipal general election. Meetings may be called by the chairman and shall be called on request of any two members, and may be held as often as necessary but not less than four times each year. Five members of the board constitute a quorum.

(b) The members of the board and of committees appointed by it shall receive no salary as members but shall be reimbursed their traveling and other expenses incurred in attending meetings of the board or committees or while in the performance of their duties as members.

(c) The secretary of the board shall be responsible for giving notice of meetings of the board and for the preparation of the agenda for meetings. He shall also be responsible for preparing and editing the minutes of the meetings of the board and the reports of the board for the municipal government and the general public.

(d) The board may appoint committees which may include employees and members of the several state, federal and local boards, commissions, departments, offices and agencies, having some competence in the matters under consideration, and may authorize such committees to make investigations and surveys and to report to the board on such matters as may be necessary to enable the board to carry out the purposes of this act.

(e) The board shall make a report of its actions taken under this act and its recommendations to the municipal executive and municipal government. The board shall make reports of its actions and recommendations to the several state boards, commissions, departments, offices and agencies as it deems advisable.

(f) All data relating to the registration of a pesticide, licensing of a pesticide applicator or issuing of a permit for pesticide use shall be public records.

SECTION 4. Registration of Pesticides

Before any pesticide may be used or sold in this ....(municipality)........ it must be registered with the board. Registration will be granted only after an application has been filed with the secretary and only after the affirmative vote of five or more of the members present at a scheduled meeting. An application for registration must include the following:

(1) Evidence that the pesticide has been registered with the United States Department of Agriculture.

(2) Reliable scientific data showing:

(a) The amount of pesticide, determinable in units of treatment concentration for specific methods of application, required to reduce pest populations to or below the economic threshold.

(b) The ecological characteristics of the pesticide in the environment, particularly its:

1. chemical stability (persistence)
2. mobility
3. solubility characteristics
4. effect on non-target organisms

(3) Such other information as the board may require.

SECTION 5. Use of Pesticides

As to each pesticide registered with the board, the board shall determine after public hearing upon notice the potential hazard, if any, to the natural resources of the ....(municipality)........ which might result from the use of such pesticide and set limitations on the proposed use that will prevent hazard to natural resources other than the target organism. If the potential hazard to the natural resources of the municipality other than the target organism is undeterminable after the hearing and the economic interests of the municipality justify cautious experimentation with the pesticide, the board may issue limited permits for use during one season in

selected areas and require that scientific data be gathered to enable the board to determine the hazard, if any, to the natural resources of the municipality other than the target organism. No pesticide shall be sold, distributed or used in the ....(municipality)..............
until regulations regarding its use are issued by the board and all applications of the pesticide shall be made in accordance with those regulations.

SECTION 6. Licenses and Permits

(1) The board shall make regulations concerning and shall issue annual licenses to qualified pesticide applicators who apply for a license. The cost of the license shall be $............. The board shall provide for the occasional inspection of work of the licensed applicators without notice to insure compliance with the statutes and regulations. Licenses of applicators who do not comply will be revoked. The board may make regulations concerning and may require and issue permits for particular types of pesticide application and for the application of particular pesticides.

(2) Any person aggrieved by any decision of the board, whether affirmative or negative in form, which relates to granting or revoking his license or permit is entitled to review thereof in accordance with the laws of the state of ....

SECTION 7. Hearing

The board shall hold with substantial adherence to the rules of evidence applicable to judicial proceedings, a public hearing relating to registration of a pesticide or to alleged or potential environmental degradation by pesticides upon the verified petition of six or more citizens filed with the board. The petition shall state the name and address of a person within the state authorized to receive service of answer and other papers in behalf of petitioners. The board shall serve a copy of the petition and notice of the hearing upon the applicant for registration or the person responsible for the alleged or potential degradation either personally or by registered mail directed to his last known post office address at least 20 days prior to the time set for the hearing which shall be held not later than 90 days from the filing of the petition. The respondent shall file his verified answer to the petition with the board and serve a copy on the person so designated by the petitioners not later than five days prior to the date set for the hearing, unless the time for answering is extended by the board for cause shown. For the purposes of any hearing under this act the secretary may issue subpoenas and administer oaths. Within 30 days after the closing of the hearing, the board shall make and file its findings of fact, conclusions of law and order, which shall be subject to review under Article .... of the ....

Law and Rules of the state of .... If the board
determines that any complaint has been filed maliciously or in bad
faith, it shall so find and the person complained against shall be
entitled to recover his expenses on the hearing in a civil action.

SECTION 8. Declaratory Ruling

(1) The board may, on petition filed with the board by any interested person, issue a declaratory ruling determining the applicability
to any person, property or state of facts of any rule or statute enforced
by it or promulgating new or amended administrative rules. Within
a reasonable time after receipt of the petition, the board shall either
deny the petition in writing or schedule the matter for hearing. If
the board denies the petition, it shall promptly notify the person who
filed the petition of its decision, including a brief statement of the
reasons therefor. If the petition is granted, full opportunity for hearing conducted under the rules of evidence applicable to a judicial
proceeding shall be afforded to interested parties. A declaratory ruling shall bind the board and all parties to the proceedings on the
statement of facts alleged unless it is altered or set aside by a court.
A ruling shall be subject to review in the .... court
of the state of .... in the manner provided for
the review of administrative decisions.

SECTION 9. Violations

(1) Any person who shall violate any of the provisions of this
ordinance or any rule, regulation or specification promulgated thereunder, shall be guilty of an offense.

(2) Each and every day during which such violation shall continue
shall constitute a separate violation and a separate offense.

SECTION 10. Penalties

Any person convicted of violating the provisions of this ordinance
or the rules, regulations and specifications promulgated thereunder
shall be fined in an amount not less than .... dollars nor more than .... for each separate offense.

SECTION 11. Existing Rights and Remedies Preserved

Nothing herein contained shall abridge or alter any rights of action
or remedies now or hereinafter existing, nor shall this ordinance, nor
any provision thereof, nor any rule or regulation promulgated thereunder, be construed as estopping the .... (municipality) ....
.... from exercising its rights and fulfilling its obligations to
protect the public health and welfare.

SECTION 12. Conflicting Laws

This article shall be construed to be ancillary to and supplementing any laws now in force tending to effect the purposes set forth in Section 1 of this ordinance, excepting as they may be in conflict herewith, in which case provisions of this Title shall govern.

SECTION 13. Local Laws, Ordinances and Regulations

Any local laws, ordinances or regulations which are not inconsistent with this article or with any code, rule or regulation which shall be promulgated pursuant to this ordinance shall not be superseded by it, and nothing in this ordinance, or in any code, rule, or regulation which shall be promulgated hereunder, shall prevent the adoption of any local laws, ordinances or regulations which are not inconsistent with this Title or with any code, rule or regulation which shall be promulgated hereunder.

SECTION 14. Separability Clause

If any clause, sentence, paragraph, section or part of this article shall be adjudged by any court of competent jurisdiction to be invalid, the judgment shall not affect, impair or invalidate the remainder of this article, but shall be confined in its operation to the clause, sentence, paragraph, section or part of this article that shall be directly involved in the controversy in which such judgment shall have been rendered.*

---

*This model law was prepared by· Victor John Yannacone, Jr. and presented to the 1969 Mid-Winter Republican Governor's Conference, Hot Springs, Arkansas.

EDF and Madison

From today's vantage point, what was the significance of the Environmental Defense Fund, its attorney Victor Yannacone, and his legal approach to solving ecological problems?

Soon after the close of the Madison hearings Yannacone and EDF severed relations, each going off in his own direction. The reasons for this were the many personal, social, and political abrasions among the trustees and financial backers. Most observers of the falling out would agree that the EDF-Yannacone combination carried within it the seeds of its own destruction. And perhaps the only thing which it could not tolerate was success.

Until the second stage of the Madison hearings, the group had almost no sense of professionalism; it was a mobile group willing to go anywhere just for the joy of taking on the big boys who were screwing up the environment. There was little money and the lack of money was replaced by that intangible—enthusiasm. Everything done by EDF was done under pressure and was loose. Thus, the small group of amateurs, who could be thrown out of one court in Michigan at 11 A.M. and file the same suit in another court two hours later, could drive terror into the hearts of a calcified DDT industry. This penny-ante approach was effective; the men had nothing really to lose. They could strike anywhere, whipping up a complaint in hours if need be.

But the stakes were suddenly raised when they got to Madison. Even though EDF had been preparing for a massive legal and scientific showdown with the DDT industry, the scientists did not realize the social implications of playing for big stakes. EDF suddenly became involved professionally up to its neck and its members could no longer be their own bosses.

The $10,000 donated by Dr. H. Lewis Batts to launch EDF's first lawsuit two years before, became a drop in the bucket. EDF wasn't in industry's league as far as finances went, but instead was dependent on massive fund-raising drives to sustain the Madison effort. And money is rarely given without some sort of string attached. Each of the groups involved with the petitioners had its own special interests and its own axe to grind whether it be EDF's choice of witnesses or Yannacone's often harsh and abrasive courtroom behavior.

EDF along with the petitioners became unwieldy at Madison. Its amateur days were over. The group suddenly had to make the transition from conservationists out to save the world and enjoy the process to a group of people whose scientific reputations were at stake. With almost every word spoken at the hearing being transmitted around the country, there was little room for error. The looseness had to go. EDF had to become tough and had to become assured of its own importance in order to bear down for the long fight.

Yannacone had his own special set of problems. He alone of all of EDF's founders was a professional in the sense that the work he performed for EDF was the type of work on which he depended for his livelihood; his courtroom work was his profession. This situation was not true for the other members of EDF's board. They were all scientists with other jobs who used spare time to testify or help out in EDF's cases. They made sacrifices but the kinds they made were qualitatively different from those which Yannacone made.

When the national spotlight turned on him, Yannacone had no laboratory to retreat to if things went badly for EDF. For the scientists, their egos were involved; for Yannacone, his professional identity. When it became known that Madison wasn't going to be a one-week frolic, this difference became crucial. Suddenly Yannacone was in the national legal big time; he became the subject of major articles in such publications as *Sports Illustrated*. He was a legal star being billed as the lawyer who wanted to save the world, and Wurster's slogan of "Sue the Bastards" became Yannacone's calling card.

There is no question that some of the other members of the EDF board became jealous of Yannacone's quickly rising star and this in some ways led to EDF's split but there was more to it than mere jealousy. The others were still amateurs and Yannacone was being billed as a professional. The others were dwarfed by him and Yannacone's often arrogant behavior capitalized on it. The small group of amateurs no longer existed.

Yannacone thought of himself as a full-time environmental lawyer and the national press encouraged this notion to the fullest. But there was one ingredient necessary to make him into this legal specialist and that of course was money. And money finally came, principally from The Ford Foundation but channeled through the Rachel Carson Fund. Yannacone began being paid his expenses and encouraged to establish a law office for environment defense. The demarcation between the amateurs and the professional was completed. But a professional surrounded by amateurs, even as battle hardened as they were by the end of the Madison hearings, spelled trouble.

Because he had accepted the challenge and opportunity to become the lawyer for the environment, Yannacone was now dependent upon EDF money to cover his mounting overhead expenses, and he was put in the position of having to take orders from those who controlled that money. Yannacone became immediately accountable to other members of EDF's board, the National Audubon Society, and the Ford Foundation, each of whom more or less demanded that he conform to their wishes. Yannacone was in the paradoxical position of giving orders during the lawsuit to all of the scientists involved and yet having to take orders from them afterwards.

This conflict between amateurs and professionals has beset many before; the outcome was inevitable and by September, 1969, the group was completely divided.

What will happen next either to Yannacone or to EDF is still some-what nebulous. But whatever occurs, that group of professors gathered around the State University of New York at Stony Brook and their country lawyer from Patchogue have left an indelible mark on the public both from an environmental and legal standpoint.

Glossary

absorption   the incorporation of a substance into another substance
adsorption   the adhesion of a substance onto another substance
anaerobic   in the absence of free oxygen
androgen   a hormone usually produced in the testes or adrenal cortex or a synthetic substance which will stimulate the development of secondary sex characteristics.
axon   a nerve cell extension that carries impulses away from the body of the cell
biphenyl   a chemical compound with the following form:

When one of the hydrogens of this compound is replaced by a chlorine, the compound is called a chlorinated biphenyl. If more than one hydrogen is replaced by chlorines, the compound is called a polychlorinated biphenyl.
carcinogen   a substance producing or inciting cancer
chlorinated hydrocarbon   a hydrocarbon (a compound of hydrogen and carbon) in which a chlorine atom occupies a position normally occupied by a hydrogen atom. Examples of chlorinated hydrocarbon insecticides follow.
aldrin

chlordane

DDD (*dichlorodiphenyldichloroethane*)

DDE (*dichlorodiphenylethane*)

DDT (*dichlorodiphenyltrichloroethane*) Two of several isomers are shown. Commercial DDT is usually composed of 15 to 20% of the *o,p* form and 80% of the *p,p* form.

Para, para—DDT   Ortho, para—DDT

dieldrin

endrin

heptachlor

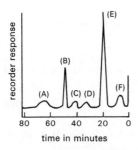

lindane

chromatography, gas a technique for separating and analyzing components of a fairly volatile mixture performed, basically, in the following manner: A column, consisting of a U-shaped glass tube containing an inert, finely divided solid, is maintained at a constant temperature in a liquid or air bath. A carrier gas is forced through the column at a desired rate and, when this rate becomes stable, a sample of the mixture to be analyzed is injected by a syringe into a vaporizing section at the front end of the column. The gas then passes through the column and, as it gradually emerges, its composition is monitored by a detecting device attached to a recorder which draws a peak whose area is proportional to the amount of the component present in the mixture. A chromatogram of a mixture of the components A, B, C, D, E, and F follows.

chromatography, thin-layer a method of separating and analyzing components of a solution. A drop of the solution to be analyzed is placed on a glass plate that has previously been coated with a thin layer of a powdered absorbing substance. The plate is then stood in a layer of a solvent which proceeds to rise up the plate. Different components of the mixture will migrate with the solvent to different positions on the glass plate.

dendrite a usually branching nerve fiber that carries impulses toward the nerve body

endocrine substances secretions produced within the body that are transported by the bloodstream to other parts of the body

endoplasmic reticulum a series of membrane enclosed channels in the cytoplasm (non-nuclear portion) of the cell, thought to function in the transport and distribution of substances between the nucleus, cytoplasm, and cell membrane and to function as a framework providing manufacturing surfaces for the cell. Smooth endoplasmic reticulum (as opposed to rough) does not have ribosomes (sites of protein synthesis) along its outer surfaces.

entomologist a scientist who studies insects

entomophagous a creature who feeds on insects. An entomophagous insect would, therefore, be an insect-eating insect.

enzyme any of a number of substances which bring about or accelerate reactions at body temperatures without being destroyed themselves: an organic catalyst

estradiol a highly estrogenic hormone found especially in the fluid of the ovary's follicles

estrogen female sex hormones found in the ovary or synthetic substances which stimulate the development of female secondary sexual characteristics and the periodic ability to conceive

food chain a sequence of organisms, each of which uses the next lower member as a food source

glucose a simple sugar which, along with water and oxygen, forms an end product of photosynthesis. It may be broken down to yield energy, or may be used as a raw material for the synthesis of other compounds such as fats, proteins, sucrose, or the storage product starch.

half-life the time required for half of a particular substance to disintegrate

hematopoiesis the formation of blood or of blood cells in a living body

histology the study of tissues

hydrolysis a chemical reaction involving the addition of a water molecule

hydroxylation the introduction of an hydroxyl group (OH) into a substance, usually in place of hydrogen

in vitro an experiment conducted on tissues or cells outside of a living body, literally "in glass" or in a test tube

in vivo an experiment conducted within a living body

ion a charged atom or group of atoms

isomers chemical compounds which contain the same number of atoms of the same elements but which differ in structure and properties. Following are the structural formulas of two isomers of DDT.

Para, para—DDT   Ortho, para—DDT

leguminous   pod-producing plants such as peas and beans, also fodder plants such as clover, alfalfa, and soybean

metabolism   the sum of the processes by which a particular substance is handled in the living body

microsome   a cell fraction obtained by centrifugation consisting of fragments of the endoplasmic reticulum (see definition above) and of ribosomes (the sites of protein synthesis in the cell)

molecular weight   the sum of the atomic weights (the average weight of the atoms of an element) of all the atoms in a molecule of a substance

mutagenesis   the induction of a mutation or basic alteration, especially in hereditary material

nonpolar solvent   a solvent made up of molecules which have no measurable separation of positive and negative charge centers. Nonpolar compounds are best dissolved by nonpolar solvents. Polar compounds, those with a measurable separation of positive and negative charge centers, are most easily dissolved in polar solvents.

oxidation   the removal of an electron from a molecule. An oxidation reaction often involves the addition of oxygen or the removal of hydrogen from the molecule.

phytotoxic   poisonous to plants

progesterone   a female sex hormone necessary to maintain pregnancy

RNA or ribonucleic acid   the hereditary material found, unlike DNA, in both nuclear and non-nuclear portions of the cell which acts as a template for protein synthesis

solution   a gas, liquid, or solid which is homogeneously mixed with another gas, liquid, or solid and will not separate under normal conditions

steroid   a complex molecule composed of four interlocking rings of carbon atoms with various side groups attached to them. Steroids, which include some hormones and some vitamins, are insoluble in water but soluble in ether.

suspension   a mixture of solids, liquids, gasses, or combinations of the three which will separate into their constituent elements unless constantly agitated

testosterone   a male sex hormone that stimulates the development and maintenance of masculine secondary sex characteristics

translocation the conduction of soluble material from one part of a plant to another, or from one location to another

trophic level the position of an organism's nutritional requirements in a larger scale of such requirements

veratrine a poisonous mixture that is a strong local irritant and muscle and nerve poison, used as a counterirritant in arthritis and as an insecticide

8 Gordon Smith from National Audubon Society

11 Black Star

15 Courtesy of Harmon Henkin

18 Courtesy of *Organic Gardening and Farming*

22 Courtesy of James Staples

23 Courtesy of James Staples

26 Courtesy of Wisconsin Department of Natural Resources

31 Elizabeth Hecker

34 Courtesy of United Nations

38-39 Michael C. T. Smith from National Audubon Society

44 Photo Researchers, Inc.

48 Courtesy of Harmon Henkin

52 Photo Researchers, Inc.

53 Courtesy of Shedd Aquarium; courtesy of U.S. Department of the Interior, Fish and Wildlife Service; Gordon S. Smith from National Audubon Society

56 Courtesy of Michigan Department of Conservation

65 Courtesy of Harmon Henkin

70 Courtesy of Harmon Henkin

74 Courtesy of Michigan Department of Conservation

76 Courtesy of Michigan Department of Conservation

80 Dave and Lyn Hancock

83 Courtesy of Edward M. Brigham

86 UPI

90 Courtesy of Harmon Henkin

105 Courtesy of Harmon Henkin

114 Elizabeth Hecker

122 Both, U.S. Department of Agriculture

123 Photo Researchers, Inc.

128 By permission of Carl Somdal, *The Chicago Tribune*

129 By permission of Conrad of the *Los Angeles Times*

136 Elizabeth Hecker

145 Courtesy of Harmon Henkin

154 U.S. Department of Agriculture

167 By permission of Johnny Hart and Field Enterprises, Inc.

177 Winston Vargas from Photo Researchers, Inc.

184 Maurits C. Escher, "Sky and Water II," from the collection of C.V.S. Roosevelt, Washington, D.C.

ABCDEFGHIJ— M —7654321